Letts

Ultimate Exam Practice

570

GCSE exam secrets

Biology

CONTENTS

To revise any of these topics more thoroughly, see *Letts Revise GCSE Biology Study Guide*

(see inside back cover for how to order)

THIS BOOK AND YOUR GCSE EXAMS

Introduction

This book is designed to help you get better results.

▶ Look at the grade A and C candidates' answers and see if you could have done better.

▶ Try the exam practice questions and then look at the answers.

▶ Make sure you understand why the answers given are correct.

▶ When you feel ready, try the GCSE mock exam papers.

If you perform well on the questions in this book you should do well in the examination. Remember that success in examinations is about hard work, not luck.

What examiners look for

▶ Examiners are obviously looking for the right answer, however it does not have to match the wording in the examiner's marking scheme exactly.

▶ Your answer will be marked correct if the answer is correct Biology even if it is not expressed exactly as it is on the mark scheme. The examiner has to use professional judgement to interpret your answers. You do not get extra marks for writing a lot of irrelevant words.

▶ You should make sure that your answer is clear, easy to read and concise.

▶ You must make sure that your diagrams are neatly drawn. You do not need to use a ruler to draw diagrams. Diagrams are often drawn too small for the examiner to see them clearly.
They should be clearly labelled with straight label lines.

▶ On many papers you will have to draw a graph. At Foundation tier axes of the graph and scales are usually given. However, at Higher tier you usually have to choose them. Make sure you always use over half the grid given and make sure you label the axes clearly. If the graph gives points that lie on a straight line use a ruler and a sharp pencil for this line. If it is curve, draw a curve rather than joining the points with a series of straight lines. Remember you may have some anomalous points and your line or curve should not go through these.

Exam technique

▶ You should spend the first few minutes of the examination looking through the whole exam paper.

▶ Use the mark allocation to guide you on how many points you need to make and how much to write.

▶ You should aim to use one minute for each mark; thus if a question has 5 marks, it should take you 5 minutes to answer the question.

▶ Plan your answers; do not write down the first thing that comes into your head. Planning is absolutely necessary in questions requiring continuous and extended answer questions.

▶ Do not plan to have time left over at the end. If you do use it usefully. Check you have answered all the questions, check arithmetic and read longer answers to make sure you have not made silly mistakes or missed things out.

DIFFERENT TYPES OF QUESTIONS

Questions on end of course examinations (called terminal examinations) are generally structured questions. Approximately 25% of the total mark, however, has to be awarded for answers requiring extended and continuous answers.

Structured questions

A structured question consists of an introduction, sometimes with a table or a diagram followed by three to six parts, each of which may be further sub-divided. The introduction provides much of the information to be used, and indicates clearly what the question is about.

▶ Make sure you read and understand the introduction before you tackle the question.

▶ Keep referring back to the introduction for clues to the answers to the questions.

Remember the examiner only includes the information that is required. For example, if you are not using data from a table in your answer you are probably not answering it correctly.
Structured questions usually start with easy parts and get harder as you go through the question. Also you do not have to get each part right before you tackle the next part.

Some questions involve calculations. Where you attempt a calculation you should always include your working. Then if you make a mistake, the examiner might still be able to give you some credit. It is also important to include units with your answer.

Some questions may be based on information from a newspaper article. Do not be alarmed if the information is new to you. You may simply have to use the information provided to answer the questions.

Some structured questions require parts to be answered with longer answers. A question requiring continuous writing needs two sentences with linked ideas. A question requiring extended writing may require an answer of six to ten lines. Many candidates taking GCSE examinations generally do less well at questions requiring extended and continuous writing. They often fail to include enough relevant scoring points and often get them in the wrong order. Marks are available on the paper here for Quality of Written Communication (QWC).

 This logo in a question shows that a mark is awarded for QWC.

To score these, you need to write in correct sentences, use appropriate scientific terminology and sequence the points in your answer correctly. Usually guidance is given about the nature of the written work being assessed, for example, 'One mark is awarded for a clear ordered answer' or 'One mark is for correct spelling, punctuation and grammar' or 'One mark is for correct use of scientific words'.

Approximately 5% of the marks are awarded for questions testing Ideas and Evidence. Usually these questions do not require you to recall knowledge. You have to use the information given to you in the question.

WHAT MAKES AN A/A*, B OR C GRADE CANDIDATE

Obviously, you want to get the highest grade that you possibly can. The way to do this is to make sure that you have a good all-round knowledge and understanding of Biology.

GRADE A* ANSWER

The specification identifies what an A, C and F candidate can do in general terms. Examiners have to interpret these criteria when they fix grade boundaries. Boundaries are not a fixed mark every year and there is not a fixed percentage who achieve a grade each year. Boundaries are fixed by looking at candidates' work and comparing the standards with candidates of previous years. If the paper is harder than usual the boundary mark will go down. The A* boundary has no criteria but is fixed initially as the same mark above the A boundary as the B boundary is below it.

GRADE A ANSWER

A grade candidates have a wide knowledge of Biology and can apply that knowledge to novel situations. An A grade candidate generally has no bad answers and scores marks throughout. An A grade candidate has to have sat Higher tier papers. The minimum percentage for an A grade candidate is about 80%.

GRADE B ANSWER

B grade candidates will have a reasonable knowledge of the topics identified as Higher tier only. The minimum percentage for a B candidate is exactly halfway between the minimum for A and C (on Higher tier).

GRADE C ANSWER

C grade candidates can get their grade either by taking Higher tier papers or by taking Foundation tier papers. There are some questions common to both papers and these are aimed at C and D candidates. The minimum percentage for a C on Foundation tier is approximately 65% but on Higher tier it is approximately 45%.

If you are likely to get a grade C or D on Higher tier you would be seriously advised to take Foundation tier papers. You will find it less stressful to get your C on Foundation tier as you will not have to answer questions targeted at A and B.

HOW TO BOOST YOUR GRADE

Grade booster ····⟩ How to turn C into B

▶ All marks have the same value. Don't forget the easy marks are just as important as the hard ones. Learn the definitions – these are easy marks in exams and they reward effort and good preparation. If you want to boost your grade, you **cannot** afford to miss out on these marks – they are easier to get.

▶ Look carefully at the command word at the start of the sentence. Make sure you understand what is required when the word is **state, suggest, describe, explain** etc.

▶ Do not confuse command words such as 'why' and 'how'.

▶ For numerical calculations, always include units.

▶ If the question asks for the name of a process, do not give a description.

▶ Read the question twice and underline or highlight key words in the questions. *e.g Suggest how <u>burning the plants</u> helped to produce <u>good crops</u> for the <u>first few years</u>.*

▶ If a lot of information is given in a question use a ruler to help you read each line.

▶ Attempt every question. If you are uncertain about an answer make an 'intelligent guess' rather than leave a blank space.

▶ Follow instructions exactly. When asked to tick **one** correct answer do exactly that. Do not tick two or more answers or draw rings, highlight or cross out.

Grade booster ····⟩ How to turn B into A/A*

▶ Check any calculations you have made at least twice, and make sure that your answer is sensible.

▶ Always write a complete answer to a question. In lessons your teacher will encourage you to add more information if you answer is too brief; this cannot take place in an examination.

▶ Try and use appropriate technical terms in your answer. This will make it concise and direct.

▶ In answering a comparison question make sure your answer is not just a description.

▶ Some questions start with the word 'suggest'. This means that there are a number of possible answers. Check that your answer is not silly or impossible and that your choice contains enough information to gain the maximum available marks.

▶ In questions requiring extended writing make sure you make enough good points and you don't miss out important points. Read the answer through and correct any spelling, punctuation and grammar mistakes.

CHAPTER 1
Cell structure and division

To revise this topic more thoroughly see Chapter 1 in *Letts Revise GCSE Biology Study Guide.*

 Try this GCSE question and then compare your answers with the Grade C and Grade A model answers on page 10.

Scientists discover two organisms and look at their cells.

cell from organism A

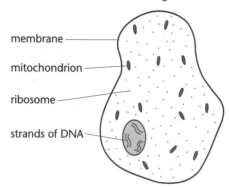

membrane
mitochondrion
ribosome
strands of DNA

cell from organism B

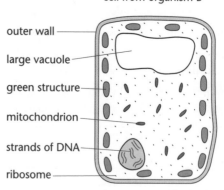

outer wall
large vacuole
green structure
mitochondrion
strands of DNA
ribosome

a   Write down the name of **one** structure which shows both cells came from living organisms and explain its function.

Name of structure ....

Function ....

.... **[2]**

b   Which cell came from an animal?
Give reasons for your answer.

....

....

....

.... **[3]**

c　Cell A was placed in different solutions.

pure water　　　　　　　　　　　　　salt solution

cell became swollen　　　　　　　　　cell shrinks

(i) Which process was involved in these changes?

Choose from,

active transport, **diffusion**, **osmosis**, **photosynthesis**.

.... **[1]**

(ii) Explain why cell A became swollen in pure water.

You will be given credit for the use of technical terms.

....

....

....

.... **[4]**

d　The scientists discovered that many of the cells from organism A had 28 chromosomes and a few cells had 14 chromosomes.

(i) Explain why these cells contained different numbers of chromosomes.

....

....

....

....

.... **[4]**

(ii) In one cell, the scientists thought there were 27 chromosomes. After recounting them they found it was really 28 chromosomes. Explain why it was unlikely that a cell would contain 27 chromosomes.

....

....

.... **[2]**

(Total 16 marks)

These two answers are at grades C and A. Compare which one your answer is closest to and think how you could have improved it.

GRADE C ANSWER

Amanda has given 2 answers. Fortunately both are correct so she gains 1 mark.
Cell metabolism is too vague an answer. Mitochondria are involved in energy release, ribosomes in protein synthesis.

Osmosis is correct – 1 mark.

Amanda correctly realises that the entry of water is important and gains 1 mark. But she does not explain (using correct technical terms) how or why it enters the cell. What eventually happens to the cell is not asked for.

Amanda

a   Both mitochondria and ribosomes ✓
      Both carry out metabolism ✗

b   Cell A ✓ It is surrounded by a membrane and not a wall ✓

c (i) Osmosis ✓

(ii) Water enters the cell, causing it to become swollen ✓ It will eventually burst ✗

d (i) The cells containing 28 chromosomes were normal body cells ✓ The cells with 14 chromosomes were sex cells ✓

(ii) Because 27 chromosomes cannot be halved ✓

Amanda has correctly identified the cell.
Only one reason is stated, yet the question asked for reasons (more than one). She only scores 2 marks out of 3.

Amanda's answers are correct. However, a more detailed answer is required to gain the extra 2 marks.

Again Amanda's answer is correct but only one fact is stated. The question asked for an explanation worth 2 marks. She gains only 1 mark.

8 marks = Grade C answer

Grade booster ···▷ move a C to a B.

The number of marks available is important. Usually there is one mark for each fact. Amanda's answers are too brief, in **c (ii)** she should have used technical terms such as semipermeable membrane, solute and solvent.

GRADE A ANSWER

Both answers correct – 2 marks. Mitochondria or ribosomes are alternative answers.

Ben has given a lot of correct information, the last answer is not marked because he has already scored the maximum 3 marks.

Osmosis is correct – 1 mark.

Ben has used technical terms such as solute, semipermeable membrane and gains the extra 2 marks.

Ben

a   Both have DNA strands ✓
      DNA carries genetic information ✓

b   Cell A ✓
      Surrounded only by a membrane and not a wall ✓ has an irregular shape ✓ does not contain green structures

c (i) Osmosis ✓

(ii) water enters by osmosis ✓ higher concentration of solute inside cell ✓ only water gets through semipermeable membrane. ✓✓

d (i) 14 chromosomes are in sex cells ✓ half the normal number of chromosomes ✓

(ii) cells must have an even number ✓ in meiosis the number is halved ✓

Cell is correctly identified – 1 mark.

Ben has not given a full answer and only scores 2 out of 4 marks

Correct explanation.

14 marks = Grade A answer

Grade booster ···▷ move A to A*

This is an excellent answer. Ben's answers were targeted at the number of marks available and he realised that extra marks were available for the use of technical terms in **c(ii)**. Only **d(i)** could be improved by explaining that the haploid (half) number of chromosomes is produced by meiosis in the production of gametes, while the diploid number is maintained by mitosis in growth.

cell structure and division

10

1

a) The following information is about cells, their adaptations and their functions.
Draw lines to match up the cell names with their adaptations and functions.
One set has been done for you.

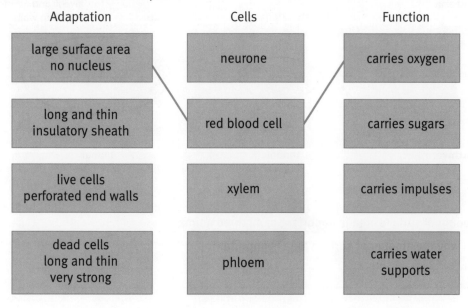

Adaptation	Cells	Function
large surface area no nucleus	neurone	carries oxygen
long and thin insulatory sheath	red blood cell	carries sugars
live cells perforated end walls	xylem	carries impulses
dead cells long and thin very strong	phloem	carries water supports

.... ④

b) This is a diagram of a neurone.

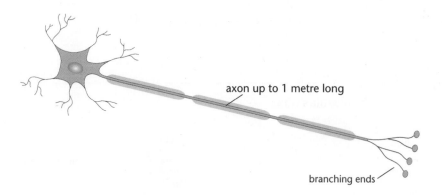

axon up to 1 metre long

branching ends

Explain the importance of the long axon and the branching ends.

....
....
.... ②

TOTAL 6

cell structure and division

2

a) All animals respire. They use up oxygen and produce carbon dioxide.

Draw a ring around the name of the process by which these gases enter and leave.

active transport,   diffusion,   osmosis,   transpiration   ①

b) The diagrams show cross-sections through different animals and how they are adapted to respire.

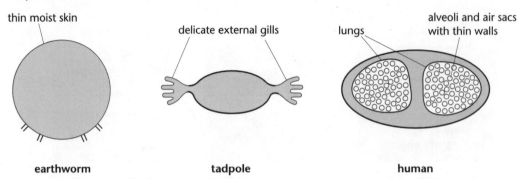

thin moist skin

delicate external gills

lungs

alveoli and air sacs with thin walls

earthworm   **tadpole**   **human**

(not to scale)

The amount of surface area is important to all three animals.

i) Explain why the amount of surface area is important.

....
....
.... ②

ii) Suggest why animals such as tadpoles and humans have special structures such as gills and lungs.

....
....
.... ②

iii) Explain why an earthworm would die if left exposed to the sun.

....
....
.... ②

c) The diagram shows the alveolus wall.

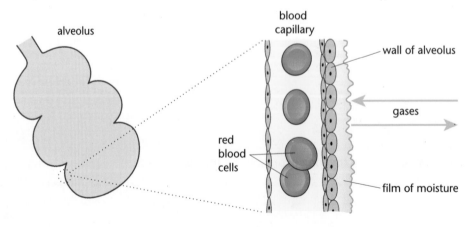

blood capillary

alveolus

wall of alveolus

gases

red blood cells

film of moisture

Explain how oxygen is absorbed.

....
....
....
.... ④

3 The two graphs show potassium and bromine ion uptake in dandelion roots.

What do these results tell you about ion uptake in plants?

....
....
....
.... ④

Cell structure and division

1 a) Long thin sheath to neurone
Live cells to phloem   all correct **2**
Dead cells to xylem
(1 mark for 1 or 2 correct)

Neurone to carries impulse
Xylem to carries water   all correct **2**
Phloem to carries sugar
(1 mark for 1 or 2 correct)

EXAMINER'S TIP

Always use a ruler so there is no doubt in the examiner's mind as to where the lines are drawn.
It is a good idea to look at all possible boxes before drawing any lines.
Always use a pencil because you may change your mind. You can then easily rub out and redraw lines without any confusion.
Join up the answers you are sure about first, this may solve any doubts you have about other answers.

b) Long axon means impulses can be carried a long way, with a direct connection between structures. Branching ends can pass impulses to many other cells
One mark for each explanation **2**

EXAMINER'S TIP

Make sure you write down which part you are explaining. If you just write down 'To pass impulses to many other cells', the examiner does not know you are referring to the branching ends.
Candidates often assume examiners are mind readers!

2 a) (diffusion) **1**

EXAMINER'S TIP

The instruction is 'Draw a ring'. Underlined, ticked or crossed-out words cause confusion and you may not be awarded the mark.

b) i) It is where oxygen enters
More surface area means more oxygen enters
For more energy release/more activity
Any two points, one mark each **2**

EXAMINER'S TIP

There are often more possible answers than the number of marks available. Make sure you give at least two pieces of information so you can gain both marks.
Candidates tend to ignore the word 'important' and not write down the reason behind getting oxygen, this is to release energy from food.

ii) More advanced or more active animals
Need more oxygen **2**

EXAMINER'S TIP

This is a difficult answer aimed at a Grade A candidate. Where the word 'Suggest' is used, it means it is not straight forward and you may need a thorough understanding of the topic.
Also there may be more than one correct answer.

iii) Skin dries or becomes hard
so oxygen cannot get through. **2**

EXAMINER'S TIP

This is another difficult question. Two marks are available, the first one is straight forward but candidates often do not complete their answer for the second mark.
Remember to look at the number of marks available.

c) During breathing in:
high concentration of oxygen in alveolus
oxygen dissolves in moisture
oxygen diffuses through walls
into blood capillary
haemoglobin in red blood cells
reacts with oxygen to form oxy-haemoglobin
Any four points, one mark each **4**

EXAMINER'S TIP

First you must realise that four marks are available. It is easy to write down a vague rambling answer and gain only 1 or 2 marks. A clear and logical answer is required.
Write down, in pencil in the margin, a list of facts you want to include, remember there must be at least four.
Then check they are in the correct order and see if you have missed out any obvious answers.
Now you can write a perfect answer!

❸ More ion uptake with oxygen
So respiration/energy involved in ion uptake
Different amounts of potassium and bromine
ions taken up
So cannot be by diffusion or osmosis, must
be active transport
Amounts of ions taken up differ between
oxygen and nitrogen
So possible different carriers more sensitive
to oxygen demand
                    Any two points followed by an
                        explanation **❹**

EXAMINER'S TIP

This is a very demanding question. You must organise your answer to include comments on the results followed by an explanation of each comment. Many candidates comment on results but do not explain them.

cell structure and division

CHAPTER 2

Humans as organisms

To revise this topic more thoroughly, see Chapter 2 in *Letts Revise GCSE Biology Study Guide.*

 Try these sample GCSE questions and then compare your answers with the Grade C and Grade A model answers on pages 18–19

1 An athlete's breathing rate is measured before, during and after a race.

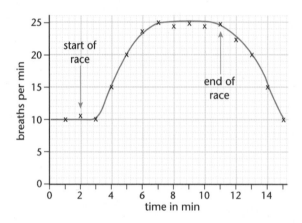

a Before the race, respiration is shown by the following word equation.

glucose + oxygen ⟶ carbon dioxide + water + energy

(i) What is this type of respiration called?

.... [1]

(ii) What does the athlete use the energy for?

....

.... [2]

(iii) Why does the athlete's breathing rate increase at the start of the race?

....

....

.... [3]

b   Towards the end of the race, a different type of respiration is used shown by the following word equation.

$$glucose \longrightarrow lactic\ acid\ +\ energy$$

(i)   Explain why this type of respiration is necessary.

....

.... **[2]**

(ii)   Explain why less energy is produced.

....

.... **[2]**

c (i)   What is the athlete's breathing rate 2 minutes after the race ended?

.... breaths/min   **[1]**

(ii)   Why is this rate different from the rate before the race?

....

.... **[3]**

2   At the start of a race, athletes react to a starter gun.

This reaction involves a number of events.

a   In the boxes below, put the following events in the correct order.

The first one has been done for you.

A   co-ordination by brain

B   detection by receptors

C   stimulus

D   response

E   effector change

C				

[3]

b (i)   Athlete's reactions can be voluntary or reflex.

Write down **one** difference between voluntary and reflex actions.

.... **[1]**

(ii)   Why are reflex actions so important to living things?

....

.... **[1]**

(Total 19 marks)

> The following two answers are at grades C and A. Compare which one your answer is closest to and think how you could have improved it.

GRADE C ANSWER

Calvin has written a poor explanation of the use of energy. This is a very common fault. The energy is used for muscle contraction and to maintain body temperature.

The first answer is correct and Calvin gains 1 mark but he wrongly believes that a lack of food is also responsible.

Calvin has been very careless in reading the information from the graph. No marks.

The last two answers are reversed so Calvin loses 1 mark.

A good answer – 1 mark.

Calvin

1a (i) aerobic respiration ✓
  (ii) to stay alive ✗
  (iii) the respiration rate increases to get more oxygen ✓
b (i) the athlete cannot get enough oxygen ✓ also not enough food ✗
  (ii) glucose is not completely broken down in anaerobic respiration ✓
c (i) 13 ✗
  (ii) he has not recovered ✓ has an oxygen debt ✓ because of lack of oxygen
2a C B A D E ✓✓ ✗
b (i) a reflex is a faster reaction, e.g. blinking ✓
  (ii) they protect from danger because of fast reactions ✓

Calvin has written down only one fact and only gains one mark.

Calvin has not completed his answer to explain about lactic acid – 1 mark only.

He has written a good answer to explain the different respiration rates but it lacks a third piece of information to gain third the mark.

A good answer, the example is not required – 1 mark.

10 marks = Grade C answer

Grade booster ····> move a C to a B

Calvin has a fair knowledge of respiration and reflex actions but he has not shown a good understanding or written many good answers. He could have avoided the vague answer of 'to stay alive' by writing down a more scientific answer. Calvin must remember that when two marks are available to explain anaerobic respiration he should try to write down two pieces of information.

When asked to get information from a graph, use a pencil and ruler to draw horizontal and vertical lines to ensure an accurate answer. It also shows the examiner where you got your information. In some situations you may get 'benefit of doubt' and get the mark.

GRADE A ANSWER

Della has been a little careless in just stating muscles. To gain the mark she should have explained that the muscles contract.

*Good answers to **b** using correct scientific words.*

A good answer. Della has accurately taken the reading from the graph.

*Della has made an unfortunate mistake. The question asks for **one** difference so the examiner only marks the first answer which does not use a comparison. Her second answer, although correct, is not marked.*

Della

1

a (i) aerobic respiration ✓

(ii) to keep body temperature constant ✓
for muscles ✗

(iii) so more energy ✓ can be released from
food ✓

b (i) lack of oxygen causes anaerobic
respiration ✓ leads to oxygen debt in the
muscles ✓

(ii) glucose has not been completely broken
down ✓ lactic acid still has some energy ✓

c (i) 20 ✓

(ii) more oxygen still needed ✓
to break down lactic acid ✓
avoids muscle fatigue ✓

2

a   C B A E D ✓✓✓

b (i) Voluntary actions involve the brain ✗
Reflex actions are faster than voluntary actions

(ii) Fast reactions help to survive from
predators ✓

A good answer, but Della has missed out the important link between energy and running a race. 'Running a race uses up more energy' would have given her the third mark.

Despite this being a difficult question, it is well answered.
Della has avoided repeating other answers she has already written down earlier in the question.

Correct answer.

Good answer.

16 marks = Grade A answer

Grade booster ┈┈➔ move A to A*

This is an excellent answer. Della has clearly shown good knowledge and understanding of respiration and reflex actions. She has also shown she can accurately extract information from a graph.

She made a common mistake of simply describing voluntary actions instead of comparing them to reflex actions.

If you are asked for one piece of information only your first answer is marked. Usually the word 'one' is printed in thick black ink. So think before you write your answer!

Humans as organisms

1 The table contains information on the internal body temperature.

Age in years	0.5	1	3	5	7	9	11	13
Internal body temperature in °C	37.6	37.5	37.2	37.0	36.8	36.7	36.6	36.5

a) i) Plot this information on the graph paper.
Draw a 'line of best fit'. **③**

ii) What is the general pattern shown by this information?

....
.... **①**

iii) Suggest an explanation for this pattern.

....
.... **①**

iv) What would you expect to find if the information continued for another 10 years?

....
.... **①**

b)   The internal body temperature is kept within a small range.

  i) What advantages does this have ?

....

....

....   ②

  ii) Explain the part played by blood vessels in the skin.
      Credit will be given for the quality of written expression.

....

....

....

....

....   ④

TOTAL 12

2

a)   The diagram shows a kidney tubule.

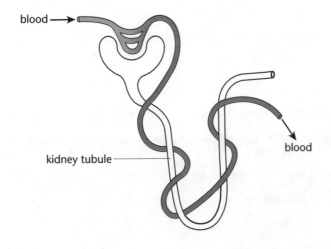

blood ⟶

blood ⟶

kidney tubule

Complete the following sentences to explain how the tubule works.

Choose from:

bladder	Bowmans	glomerulus	glucose	kidney
nephron	neurone	osmosis	ultra-filtration	urea

A kidney tubule is called a ....

Blood capillaries are surrounded by the .... capsule.

The dissolved chemicals in the blood are forced out by .... and
travel down the tubule.

Useful chemicals such as .... are re-absorbed back into the blood.

The remaining chemicals are carried to the ....   ⑤

b) The pituitary gland produces a hormone which controls the amount of water in the blood.

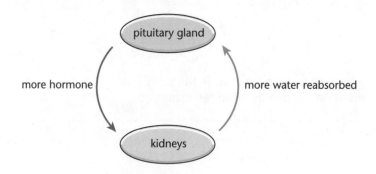

i) What is the name of this hormone? .... ①

ii) What effect would more of this hormone being produced have on the urine?

.... ①

iii) Suggest why the body would need to re-absorb more water.

.... ①

iv) Why are hormonal reactions generally slower than nervous reactions?

....

.... ②

c) The table contains information on the cost of different types of treatment for kidney disease.

Method	Cost in £s in first year	Cost in £s in each following year
A successful transplant	14 500	1 100
B blood dialysis at patient's home	15 500	6 000
C blood dialysis in hospital	13 500	13 500
D dialysis of other body fluids in hospital	8 000	7 000

i) Which method A, B, C or D would cost most in the first year of treatment? .... ①

ii) Which method A, B, C or D would cost the least over a total of five years?

Answer .... ①

iii) Suggest **two** reasons, apart from cost, why some people needing a kidney transplant do not get one.

1 ....

2 .... ②

TOTAL 14

3   Look at the diagram of the heart.

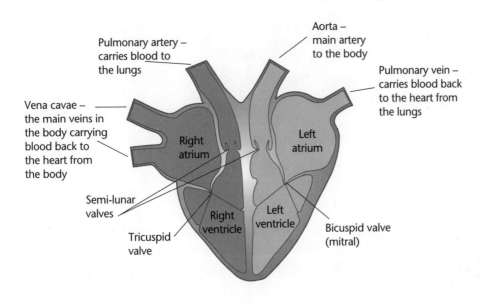

a)   The heart has a wall separating the right and left sides of the heart.
     Why is this wall necessary?
     ....   ①

b)   The heart has two atria and two ventricles.
     What are the main functions of atria and ventricles?

     Atria ....

     Ventricles ....   ②

c)   Draw arrows on the diagram to show the direction of blood flow in the **left** side of the
     heart.   ②

d)   Explain why the left ventricle has a very thick wall.
     ....
     ....   ②

e)   Damaged heart valves can be replaced.

  i) Why are heart valves necessary?
      ....   ①

  ii) Where else in the circulatory system are valves necessary?
      ....   ①

  iii) Valve replacement surgery can be risky.
       Suggest why.
       ....
       ....
       ....   ③

                                                                                                                                              TOTAL 12

4 Peter is a mountain climber.
He goes on a climbing expedition.
The number of red blood cells he has is measured.

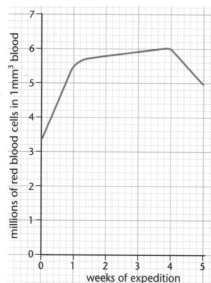

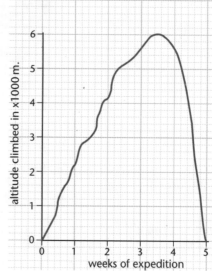

a) How high did he climb?

.... m ①

b) i) What general pattern is shown linking numbers of red blood cells to altitude climbed?

....

.... ①

ii) Suggest an explanation for this pattern.

....

....

.... ③

c) i) How many red blood cells were present in 1 mm³ blood:

at the start of the expedition ....

at the end of the expedition .... ? ②

ii) Use this information to explain why athletes train at high altitudes before a race.

....

....

.... ④

TOTAL 11

❶ a) i) Your graph should look like this.

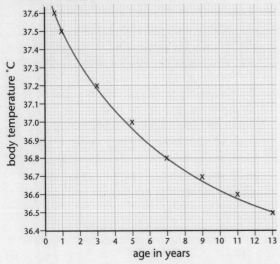

1 mark for axes, 1 mark for plotting,
1 mark for line of best fit ❸

EXAMINER'S TIP

This question would be on the Higher Tier paper since you have to decide on suitable axes and scale. In scaling, your graph should be more than half of the space provided. In graphs like this, the horizontal scale is used for fixed data, the vertical axis for measured data.
A sharp pencil is a must! It is easy to make an error in plotting but a pencil mark is easily rubbed out. Use an X to mark each point. Examiners allow only a small tolerance in plotting so do it carefully.
A line of best fit will show the general trend such as a straight line or a simple curve. If it appears to zigzag, check your plotting. Unless an anomalous plot has been used, it is probably wrong!
A word of caution; graph plotting is worth only 3 marks so learn to plot graphs quickly and accurately.

ii) Temperature decreases as age increases ❶
iii) When young there is a high growth rate requiring a high metabolic rate. This growth rate slows down with age ❶
iv) Temperature becomes constant/even/stable ❶

EXAMINER'S TIP

An explanation of a pattern should include what happens to the two variables. If you have simply written, 'Temperature increases', this is not a pattern.
The explanation of this pattern is difficult. It is an A* question.
If you continue the curve on the graph you see it will level out, meaning the temperature levels out.

b) i) Metabolism not affected by outside temperatures
Allows enzymes to work at constant rate
Enzymes not denatured by high temperatures
Enzymes work at their optimum temperatures
any two points, 1 mark each ❷

EXAMINER'S TIP

Since the word 'advantages' is used you must write down more than one. This is another difficult question targeted at Grade A*. If you have written a good answer, well done!
Remember that enzymes are denatured NOT killed.

ii) Too hot so blood capillaries in skin dilate ❶
Too cold so blood capillaries in skin close down ❶
Linking of explanation, e.g. dilated blood capillaries cause more blood to flow near skin surface ❶
Use of scientific terms, e.g. vasodilation, vasoconstriction ❶

EXAMINERS TIP

This type of question is new. In this case two marks are available for the quality of your answer. It is therefore very important to make sure your answer is written very clearly and you use scientific words.
Did you make the common mistake of writing about blood capillaries moving up or down in the skin?
The blood capillaries dilate or constrict, they do NOT move.
Many candidates make this mistake and get no marks.

Humans as organisms

❶ a) nephron
Bowmans
ultra-filtration
glucose
bladder

        1 mark for each correct answer ❺

EXAMINER'S TIP

This type of question is common in the Foundation Tier. There are always some words called 'distracters' to tempt you into the wrong answer so think carefully and look at all the possible words. Remember, you cannot make up your own words!

b) i) Antidiuretic hormone/ADH   ❶
  ii) More concentrated urine   ❶
  iii) Too much water lost/shortage of water
     taken in   ❶

EXAMINER'S TIP

ADH is an accepted abbreviation for antidiuretic hormone. However, never make up your own abbreviations to save time. The examiner cannot mind read!

  iv) Hormones are carried in the blood but nerve impulses are carried by special neurones
    Hormones result in chemical changes, nerve impulses are mainly electrical
    Hormones usually have a general action, nerve impulses are carried direct to a target organ or cells

       any two points, one mark each ❷

EXAMINER'S TIP

Does your answer have a comparison or is it just a description? Descriptions gain no marks. If you were asked for a comparison between a cat and a dog, it is no good simply to say, 'A dog barks', without saying what a cat does!

c) i) B   ❶
  ii) A   ❶
  iii) Kidneys not available/not enough donors
    Right tissue match not available
    Patient too ill for surgery any two points ❷

EXAMINER'S TIP

To answer ii) seems a lot of work but simple mathematical skills such as this are expected in a Grade C candidate. You are allowed a calculator so remember to take one into the examination and use it.
You have to work out each method,
A 1st year is £14 500 + (4 × £1100)   = £18 900
B 1st year is £15 500 + (4 × £6000)   = £39 500
C 1st year is £13 500 + (4 × £13 500) = £67 500
D 1st year is £8000 + (4 × £7000) + £36 000
So the cheapest is A.
It is quite common for candidates to carry out a mass of calculations but then forget to write down the answer. So keep concentrating.
In iii) your answer cannot mention anything to do with money, e.g. not enough nurses/hospitals.

❸ a) To separate oxygenated and deoxygenated blood   ❶
 b) To collect blood
    To send out blood   ❷
 c) Arrow entering left atrium
    Arrow entering left ventricle
    Arrow leaving through aorta
          at least two arrows ❷
 d) To force/pump blood
    All around the body   ❷
 e) i) So blood only flows one way   ❶
  ii) In veins   ❶
  iii) Heart stopped/blood diverted
    Blood carries oxygen
    So these have to be done by machine
    (accept rejection for one mark)   ❸

❹ a) 6000   ❶
 b) i) The higher he climbs, the more red blood cells are produced   ❶

EXAMINER'S TIP

There are always two parts to a pattern. You must include both parts for the mark.

  ii) More oxygen/food are needed
    When climbing/less oxygen at higher altitudes
    So body produces more red blood cells ❸
 c) i) 3.4 million
    5 million   units must be shown, if not deduct one mark ❷
  ii) More red blood cells produced
    Can carry more oxygen in race
    More energy released
    Can run faster/perform better   ❹

CHAPTER 3

Green plants as organisms

To revise this topic more thoroughly, see Chapter 3 in *Letts Revise GCSE Biology Study Guide.*

 Try this sample GCSE question then compare your answers with the Grade C and Grade A model answers on page 29.

Space travel will involve journeys lasting many years.

Scientists are looking at ways of growing food in space.

Small plants called algae could be used.

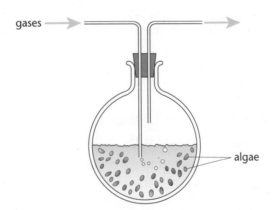

gases

algae

a (i) Name the **two** gases the plants would need and for each gas, state the process in which it is used.

....

.... [2]

(ii) What other conditions would the algae need to make food?

....

....

.... [3]

(iii) Apart from providing food, how may the algae be useful to space travellers?

....

.... [2]

b Scientists looked at the cells of two different algae.

Each cell contained green chloroplasts.

alga A alga B

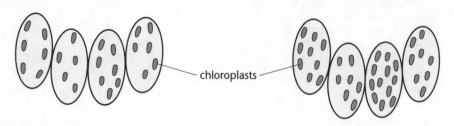

chloroplasts

(i) Complete the table below.

Algae	Total number of chloroplasts in all cells	Average number of chloroplasts per cell
alga A		
alga B		

[4]

(ii) Which alga will produce the most food?

Explain your answer.

....

....

.... **[2]**

c The scientists tried to test the cells for sugars.

They found very little sugars present in the cells.

(i) Suggest what could have happened to the sugars.

....

....

.... **[2]**

(ii) Apart from carbohydrates, what other types of food are made by plants?

.... **[2]**

d The scientists thought that if the excretory wastes of the space travellers were added to the algae, they would grow more quickly.

(i) Suggest **one** reason why this might work.

.... **[1]**

(ii) Suggest one reason why this might not work.

.... **[1]**

(Total 19 marks)

These two answers are at grades C and A. Compare which one your answer is closest to and think how you could have improved it.

GRADE C ANSWER

Eva has been a little careless and not written down that oxygen is used for respiration. Since there are only two marks for this question she must get both parts right for a mark.

Eva has written a good answer – 2 marks.

Eva not has counted the chloroplasts correctly. She only loses one mark since examiners use the idea 'error carried forward' in calculations. Based on her calculation, the second answer is acceptable.

Eva did not attempt this question.

Eva

a (i) oxygen ✗ carbon dioxide for
        photosynthesis ✓
   (ii) light ✓   temperature ✗
  (iii) give out oxygen ✓   so travellers can
        respire ✓
b (i) 25 × 6.25 ✓   36 ✓   9 ✓
   (ii) alga B more chloroplasts ✓
c (i) used it for energy ✓
   (ii) proteins ✓
d (i) ....   ✗
   (ii) cause disease and kill algae ✓

11 marks = Grade C answer

Eva has shown a common mistake in writing 'temperature'. This answer should be qualified, such as a suitable temperature or a figure such as 25℃.

Since there are only two choices, examiners do not give a mark for naming the alga. Both marks are for the explanation and Eva only writes down one piece of information.

Eva again has only written down one correct piece of information for both answers – only 1 mark for each.

A good answer – 1 mark.

Grade booster ···⟩ move a C to a B

Questions which follow a story are sometimes used. Do not panic if you know nothing about space travel, it doesn't matter. By checking her counting of the chloroplasts, she would have probably realised her first answer was wrong. When counting information like this, use a pencil to cross out chloroplasts as you count them. Never leave a blank space for an answer. Later in the examination, return to any questions you have not attempted and try again. You may now be inspired! If not have a guess, you may be lucky.

GRADE A ANSWER

Fokrul has correctly named two gases and stated why each is needed – 2 marks.

Fokrul has given a good answer, stating an actual temperature – 2 marks.

Fokrul has made a silly but common mistake, even for a grade A candidate. He has forgotten to write down which alga he has selected.

Fokrul

a (i) oxygen for respiration ✓ carbon dioxide
        for photosynthesis ✓
   (ii) light ✓ warm temperature of 25℃ ✓
  (iii) uses up the waste carbon dioxide from
        travellers ✓
b (i) 24 ✓   6 ✓
        36 ✓   9 ✓
   (ii) more chloroplasts to get more light ✗
c (i) made into starch for storage ✓
        or used as energy for growth ✓
   (ii) proteins ✓ fats and oils ✓
d (i) gives algae minerals such as nitrogen ✓
   (ii) may cause disease ✓

15 marks = Grade A answer

Although writing one correct answer, Fokrul has not realised that there are two marks for this question.

All the figures are correct – 4 marks.

All these answers are correct with full marks.

All correct in **d** – 2 marks.

Grade booster ···⟩ move A to A*

A very good set of answers. Fokrul has made a simple mistake in not naming the alga.

In one question he wrote down only one answer when two marks were on offer. If he had avoided these two simple mistakes he would have got full marks, a Grade A*.

Green plants as organisms

Green plants as organisms

1 A class looks at a glass tank.
It contains plants and is kept in bright sunlight.

bubbles of gas

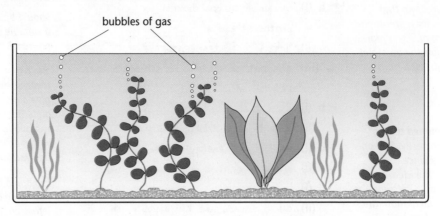

a) They see bubbles of gas coming from some plants.
What is the name of the gas being given off?

.... ①

b) The number of gas bubbles from one plant during the day were counted.
This data is shown in the graph.

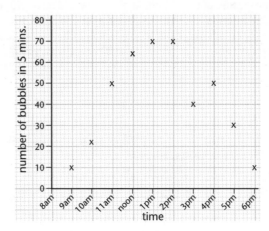

i) Using the plotted points, draw a 'line of best fit'. ①

ii) Explain the shape of the graph during the morning.

....

....

.... ③

iii) Explain the data for 3 pm.

.... ①

c) Complete and balance the equation for photosynthesis.

$$\ldots\ldots CO_2 + \ldots\ldots H_2O \longrightarrow C_6H_{12}O_6 + \ldots\ldots \ldots\ldots$$

③

TOTAL 9

2 The diagram shows the inside of a leaf.

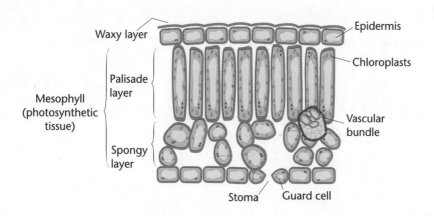

a) Name **two** types of cell layers which carry out photosynthesis.

.... ②

b) Which type of cell will provide the most food ?
Give reasons for your answer.

....
....
.... ③

c) Leaf miners are small insects which burrow into leaves and eat cells.
They make a white or transparent line in the leaves.
Suggest an explanation for this.

....
....
.... ②

d) Name the part of the leaf which carries water and food.

.... ①

e) Leaves lose water through their leaves.
Explain in detail how this happens.
One mark is awarded for a clear ordered answer and one mark is for the correct use
of scientific terms.

....
....
....
....
....
.... ⑤

TOTAL 13

3 Water is carried up plant stems by:
A transpiration pull
B root pressure
C capillarity

a) By which process A, B or C is most water carried up stems in the summer?
Give a reason for your answer.

....

....

.... ②

b) Suggest why very little water is carried up stems in winter.

....

....

.... ②

TOTAL 4

4 The tips of young plants were treated in different ways.

A
tip cut off

B
light from
the right

C
tip cut off
and replaced
on plastic

D
tip covered
in black plastic

E
light from
above

a) After a few days which stem, A, B, C, D or E,

i) grew straight upward ....

ii) did not grow at all ....

iii) curved to the right. .... ③

b) What is the reaction called when a plant curves towards light?
Choose from,

hydrotropism   geotropism   phototropism   reflex

.... ①

c) What is the name of the chemicals which cause these reactions?

.... ①

d) Explain how these experiments could show:

i) where the plant hormones were produced

ii) how they were carried round the plant.

....

....

....

....
④

TOTAL 9

5 Eva looked at a tree in her garden.
When it was first planted it was tied tightly by wire, to a post.

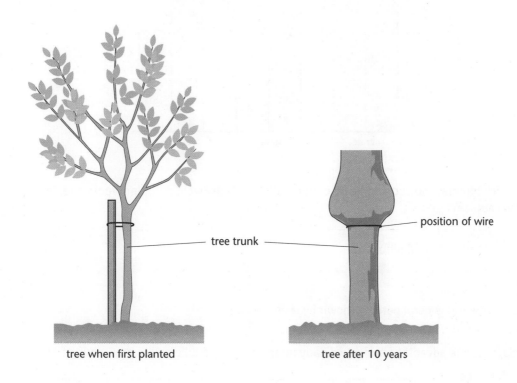

tree trunk

position of wire

tree when first planted

tree after 10 years

a) Describe the changes the wire caused to the tree trunk after 10 years.

....

....

....
②

b) Suggest explanations for these changes.

....

....

....

....
④

c) Suggest what might happen to the tree in another 10 years.

....

....
①

TOTAL 7

6 Scientists have found that many fungi living with plant roots help plant roots.
These fungi are called mycorrhiza.
The bar graph shows how mineral uptake is affected by these fungi.

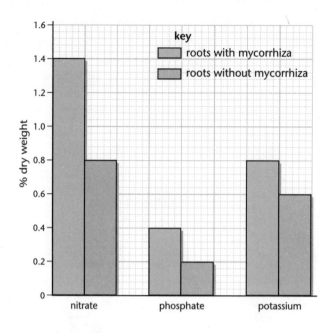

a) i) Which mineral showed an increase from 0.8 to 1.4% dry weight in plant roots affected by mycorrhiza?

.... ①

ii) Why do plants need nitrogen?

....
.... ②

b) Describe the general pattern shown by these results.

.... ①

c) Suggest **one** advantage and **one** disadvantage of using sterile soil to grow young trees.

Advantage ....
....

Disadvantage ....
.... ③

d) Suggest what benefit the fungus may gain from this relationship.

....
.... ①

TOTAL 8

❶ a) oxygen ❶

EXAMINER'S TIP

Although plants also carry out respiration, producing carbon dioxide, in bright sunlight photosynthesis will be the major process. Photosynthesis produces oxygen.

b) i) Your graph should look like this with a line drawn as a curve, missing out the data for 3 pm. ❶

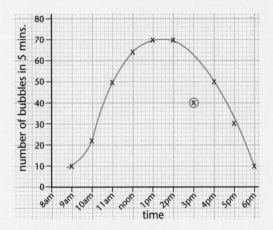

EXAMINER'S TIP

A sharp pencil is a must for you to draw a smooth curve. Did you realise that the data for 3 pm did not fit the general pattern? Your curve should have ignored this plot.

ii) Number of bubbles increased during the morning
Because of more light
Results in more photosynthesis
Which gives out more oxygen
        any two points for two marks ❸
iii) Anomalous data/data misread/bubbles miscounted/sun may have 'gone in' ❶

EXAMINER'S TIP

Did you read all of this question before answering? If you did you would have realised when you were drawing the graph that this point was wrong.
Reading the whole question first can be useful.

c) $6CO_2 + 6H_2O \longrightarrow C_6H_{12}O_6 + 6O_2$
        1 mark for each part ❸

EXAMINER'S TIP

Unlike Chemistry, you have to remember only ONE equation, so make a special effort to remember all these 6s. The equation for respiration is exactly the same as this one except it is reversed.

❷ a) i) Palisade layer
Spongy layer (often called spongy mesophyll) ❷

EXAMINER'S TIP

Although guard cells also contain chloroplasts they are not in layers so they are not an acceptable answer.

b) Cells in palisade layer ❶
Because they contain most chloroplasts
(or get more light from above)
So more photosynthesis
        any two points for 2 marks
c) Eat cells containing green chloroplasts
leaving no colour ❷
d) Vascular bundle ❶
(answers such as vein, midrib, xylem, phloem are acceptable)
e) Water inside leaf evaporates
Water vapour spreads out
Water vapour escapes through holes in lower surface of the leaf ❸
correct sequence of events ❶
use of scientific words such as, stoma, diffusion, transpiration ❶

EXAMINER'S TIP

Where marks are available for 'the quality of written expression', you must make a special effort to write your answer in good English. Your answer should put events in the right order and use the correct scientific words. To get all this right, plan your answer in pencil in the margin or on a spare page.

Green plants as organisms

❸ i) A transpiration pull
More leaves, therefore more water loss
Or
Hot weather results in faster transpiration ❷
ii) No leaves
No transpiration ❷
Or
No photosynthesis
So less water required

❹ a) D or E
A or C
B ❸
b) Phototropism ❶
c) Plant hormones/auxins ❶
d i) Produced at tips/apices
A does not curve and B does ❷
ii) Hormone is carried in water
If C does not grow upright ❷

EXAMINER'S TIP

This topic is often badly done in examinations, so make sure YOU know how plants react.

❺ a) Tree trunk thicker above the wire after 10 years
Bulge just above the wire after 10 years ❷
b) Growth
In height/thickness
Bulge caused by food not being able to get past restriction
So less growth below wire ❹
c) top heavy/collapse/easily blown down ❶

❻ a) i) Nitrogen ❶
ii) Growth
Proteins/enzymes ❷
b) Mycorrhiza increases mineral uptake ❶
c) Advantage; no diseases/pests ❶
Disadvantage; no fungi/mycorrhiza
less minerals taken up ❷
d) Food/water ❶

CHAPTER 4
Variation, inheritance and evolution

To revise this topic more thoroughly, see Chapter 4 in *Letts Revise GCSE Biology Study Guide.*

 Try these sample GCSE questions and then compare your answers with the Grade C and Grade A model answers on page 39.

1

a Athletes show a mixture of environmental and genetic characteristics.

Natural black hair

Sun tan

Long legs

Blue eyes

Tattoo

Powerful muscles

Sort out these characteristics into the table below.

Environmental characteristics	Genetic characteristics
1	1
2	2
3	3

[3]

b Charles Darwin believed that all living things show differences in characteristics resulting in variation. Those organisms having characteristics best suited to their environment would survive.

Jean-Baptiste Lamarck believed that living organisms changed their characteristics through their lifetime.

Giraffes have long necks and eat leaves from tall trees.

(i) Using Charles Darwin's ideas, explain why giraffes now have long necks.

....

....

....

[3]

(ii) Using Jean-Baptiste Lamarck's ideas, explain why giraffes now have long necks.

....

.... **[2]**

c Variation can be the result of mutations.

 (i) What is a mutation?

.... **[2]**

 (ii) State **two** causes of mutation.

 1 ....

 2 .... **[2]**

d Some conditions in humans are caused by faulty genes and chromosomes.

In the boxes, tick **two** examples of these conditions.

Down's syndrome	
Aids	
Cancer	
Cystic fibrosis	
Plague	

[2]

2 Complete the following sentences.

Choose from:

> **bond   cytoplasm   DNA   enzymes   genes   helix   IAA   nucleus**

Chromosomes are found in the cell's ....

Chromosomes carry genetic information in sections called ....

Genes are made of a chemical called ....

This chemical is in the form of a double .... **[4]**

3 In sexual reproduction, gametes are formed.

Male and female gametes fuse to form a fertilised egg.

Complete the diagram to show chromosome numbers in this process in humans.

[6]

(Total 24 marks)

These two answers are at Grades C and A. Compare which one your answer is closest to and and think how you could have improved it.

GRADE C ANSWER

George

1

a   sun tan ✓
     tattoo ✓   blue eyes ✓
     powerful muscles ✓   long legs ✓
     natural black hair ✗

b (i) giraffes can eat leaves off tall trees ✗
      those with longer necks will live longer ✓
  (ii) giraffe's necks grow longer ✓

c (i) problems with DNA ✓
  (ii) X-rays ✓   chemicals ✗

d   Down's syndrome ✓   Aids ✗

2   nucleus ✓ genes ✓ IAA ✗ Bond ✗

3   46   .... ✗
     23 ✓   23 ✓   haploid ✓
             46 ✓   haploid ✗

13 marks = Grade C answer

One wrong answer gives 2 marks out of 3. George has not realised the significance of the word 'natural' and thinks the hair colour has been changed. He has also ignored the space for three answers and written four, something is wrong!

Again, George has not written a full explanation. George has gained one mark but has not realised that it is a random change. Did you?

Although some chemicals such as mustard gas cause mutations, the answer is too vague considering the thousands of known chemicals – only 1 mark.

This just repeats the words in the question. George has not finished his answer, which is worth 3 marks.

George has picked Aids, probably because it is simply a well known condition – only 1 mark.

George has been tempted by the wrong initials IAA which is a plant hormone. He has also mixed ideas of a chemical double bond. Two marks out of 4.

George has not given an answer for female body cells, probably a careless mistake rather than not knowing. After fertilisation, the diploid number returns. 4 marks out of 6.

Grade booster ⇢ move a C to a B

George has made a few careless mistakes and put a Grade C in doubt. He left one answer blank instead of having a guess, and written four answers when space for only three was supplied. In questions which ask you to supply the missing word, usually two or three alternatives for each gap are supplied.

GRADE A ANSWER

HAZEL

1

a   suntan ✓   blue eyes ✓
     tattoo ✓   long legs ✓
     powerful muscles ✓   natural black hair ✓

b (i) giraffes' necks show variation ✓
      those with longer necks survive and breed ✓
      offspring will have taller necks ✓
  (ii) giraffes' necks grow during life ✓

c (i) a random change ✓   in chromosome number ✓
  (ii) radiation ✓   infra red light ✗

d   Down's syndrome ✓   cystic fibrosis ✓

2   nucleus ✓   genes ✓   IAA ✗   helix ✓

3   46   46 ✓
     23 ✓   23 ✓   haploid ✓
             46 ✓   diploid ✓

21 marks = Grade A answer

All answers are correct so 3 marks. One mark would have been deducted for each wrong answer.

Hazel hasn't completed her answer by writing that the offspring will then have long necks – 1 mark out of 2.

Hazel has also been tempted to use IAA as an answer. She knows initials are involved but has chosen the wrong ones! She loses 1 mark.

A good answer for three marks.

A good answer, mutations also occur in DNA.

Hazel has confused infra red with ultra violet light. 1 mark out of 2

Both answers are correct – 2 marks.

All correct – 6 marks.

Examiner's comment

Hazel's answers were excellent. Were yours? She thoroughly understands the principles behind the two different theories of evolution. This is a new type of question. It examines Ideas and Evidence in Science and is called Sc1.

Variation, inheritance and evolution

Variation, inheritance and evolution

1 The diagram contains information on the evolution of the horse.

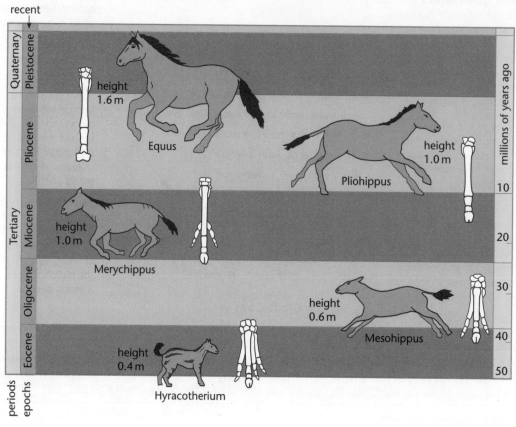

a) i) One ancestor of the horse was only 0.4 m high.
   How long ago did it live?

   .... ①

   ii) During which epoch did it live?

   .... ①

b) Describe the changes of a horse's leg length and number of toes during time.

   ....
   ....
   ....
   .... ④

c) Using Darwin's ideas on evolution, explain the changes in horse size.
   One mark is awarded for a clear ordered answer and one mark is for the correct use
   of scientific terms.

   ....
   ....
   ....
   ....
   .... ⑤

TOTAL 11

2 The diagram shows the sex chromosomes of a human couple, A and B, and four of their children, C, D, E and F.

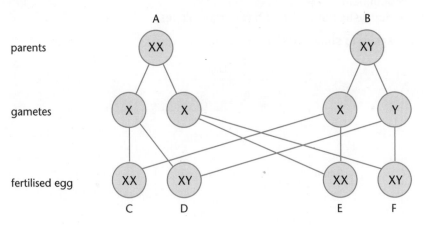

a) i) Which parent is the mother?

.... ①

ii) Explain why you chose this parent.

.... ①

b) Another child is expected. What are the chances of it being a girl ?

.... ①

TOTAL 3

---

3 Red-green colour blindness is a sex-linked inherited characteristic.
The gene (allele) **b** colour blindness is recessive to the gene **B** for normal vision.
The gene **b** is carried only on the X chromosome.

a) Describe these phenotypes.

b) Explain why red-green colour blindness is more likely to occur in men than in women.

.... ②

TOTAL 8

4 Cystic fibrosis is a genetic disease.
It affects the lungs and digestive system. It is caused by a recessive allele (r), the normal condition (R) is dominant.
Mum is a carrier for cystic fibrosis, dad has normal alleles.

a) Complete the diagram to show possible children.

Dad

	R	R
	R r	

Mum

⑤

b) Will any of their children have cystic fibrosis?
Explain your answer.

....
.... ②

TOTAL 7

5 Neil has a pig farm. Pigs are bred to produce meat.

Neil decides to improve his herd by using selective breeding.

a) Describe how this is done.

....
....
.... ③

b) Describe a different way, apart from selective breeding, in which Neil could improve his herd of pigs.

....
....
.... ③

TOTAL 6

1 a) i) Any figure between 40 and 50 million years ago. ❶
   ii) Eocene ❶

EXAMINER'S TIP

These two questions are called data interpretation questions. Always look very carefully at all the information before you answer.

Did you forget to write 'million years'? If so you would lose the mark for your answer.

b) Leg length becomes longer
   As horse becomes bigger in more recent times ❷
   Number of toes reduce from 4 to 1
   In recent times ❷

EXAMINER'S TIP

These two questions are trying to find out if you have understood the data in the diagram. Many candidates write rambling answers to this sort of question. You must make it clear when you are writing about leg length and when about number of toes. It is a good idea to first write some facts down in pencil in the margin so you can write a well organised answer. The question referred to 'during time', so this must be made clear in your answer.

c) Variation in size of early horses such as Hyracotherium
   Largest would survive and breed
   This is called Natural Selection
   Offspring would be larger
   Cycle repeated over many generations
      one mark for each point, maximum of ❸
      plus ❷ for a well written answer

EXAMINER'S TIP

Where marks are available for 'the quality of written expression', you must make a special effort to write your answer in good English. Since there are a number of points you could mention, the examiner will be looking for a good logical answer linking up the points you mention.

2 a) i) Parent A ❶
   ii) Has XX chromosomes or both sex chromosomes are the same ❶
   b) 50/50 or equal or 1 to 1 ❶

EXAMINER'S TIP

All chromosomes in a nucleus are paired off. One pair, called the sex chromosomes, determine sex. In the female, both chromosomes are the same (XX), in the male they are different (XY). When gametes fuse in sexual reproduction, there is always an equal chance of the offspring being male or female, even after four boys!

3 a) Female
   Normal vision ❷
   Male
   Red green colour blind ❷
   Male
   Normal vision ❷

EXAMINER'S TIP

This is a difficult question. There are two marks for each part so two facts are required for each part. Did you realise that 'phenotypes' meant what characteristics they had and showed?

b) In male, only one allele from a pair is present
   So if recessive factor **b** is present it will show as red-green colour blindness ❷

4 a)   r   (Rr)   Rr
         R   RR   RR ❺

EXAMINER'S TIP

Many candidates have difficulty with this type of genetics problem so make sure you have plenty of practice.

Remember that a carrier will have the 'hidden' recessive factor r present but not expressed.

b) None
   Because no rr/ no double recessive ❷

5 a) Select best characteristics
   Such as size/weight/muscle
   Breed from these parents ❸

b) Bring in new and better/bigger/ faster-growing pigs
   To breed with original pigs
   This gives better gene pool of characteristics
   Increases variation to select from
      any three points ❸
   Good answer explaining genetic engineering for 3 marks.

EXAMINER'S TIP

In 5b other answers, such as improved diet, are acceptable. However, it would be difficult to gain 3 marks from this answer. Where there are a number of different ways of answering a question, make sure the one you select can be awarded the maximum marks.

Variation, inheritance and evolution

To revise this topic more thoroughly, see Chapter 5 in *Letts Revise GCSE Biology Study Guide.*

 Try these sample GCSE questions and then compare your answers with the Grade C and Grade A model answers on pages 46–47.

1 The table and bar graph contain information about different types of household rubbish in Britain in 1960 and 2000.

Material	Percentage in 1960	Percentage in 2000
dust and ash	45	
paper and cardboard	25	
plastic	5	
glass	10	
waste food	15	

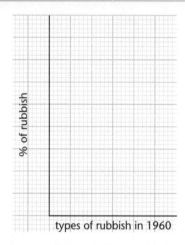

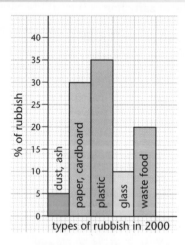

a (i) Draw a bar graph to show the information for 1960. **[2]**

(ii) Using the bar graph for the year 2000, complete the table. **[2]**

b Which material formed the largest percentage of waste in:

1960 ....

2000 .... ? **[2]**

c Describe the change in the percentage of dust and ash in the rubbish from 1960 to 2000 and suggest reasons for the change.

.... [3]

d State **one** reason, apart from cost, why it is important for paper to be recycled.

.... [1]

2 The diagram shows a food chain in the sea.

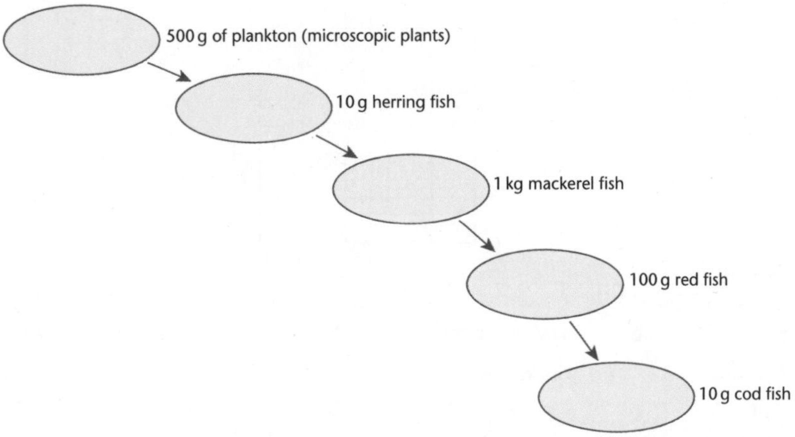

500 g of plankton (microscopic plants)

10 g herring fish

1 kg mackerel fish

100 g red fish

10 g cod fish

a Name
   (i) a producer ....
   (ii) a consumer. .... [1]

b What is the source of energy for this food chain? .... [1]

c Draw a diagram to show a pyramid of biomass for this food chain.
   An accurate scale diagram is **not** required. [2]

d Explain why 1 kg of mackerel is produced compared to only 10 g of cod.

....

.... [2]

e Neil and Anita decide to set up a fish farm. Which fish from the above food chain should they use?
   Give reasons for your choice.

....

.... [3]

(Total 19 marks)

> The following two answers are at Grades C and A. Compare which one your answer is closest to and think how you could have improved it.

GRADE C ANSWER

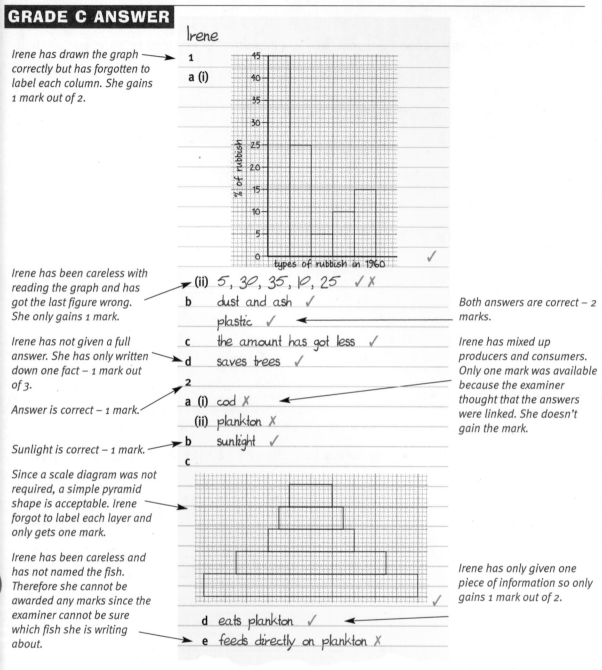

Irene

1 a (i)

% of rubbish / types of rubbish in 1960 ✓

Irene has drawn the graph correctly but has forgotten to label each column. She gains 1 mark out of 2.

Irene has been careless with reading the graph and has got the last figure wrong. She only gains 1 mark.

(ii) 5, 30, 35, 10, 25 ✓ ✗

b dust and ash ✓
   plastic ✓

Both answers are correct – 2 marks.

c the amount has got less ✓

Irene has not given a full answer. She has only written down one fact – 1 mark out of 3.

d saves trees ✓

Answer is correct – 1 mark.

2
a (i) cod ✗
   (ii) plankton ✗

Irene has mixed up producers and consumers. Only one mark was available because the examiner thought that the answers were linked. She doesn't gain the mark.

Sunlight is correct – 1 mark.

b sunlight ✓

c

Since a scale diagram was not required, a simple pyramid shape is acceptable. Irene forgot to label each layer and only gets one mark.

Irene has been careless and has not named the fish. Therefore she cannot be awarded any marks since the examiner cannot be sure which fish she is writing about.

d eats plankton ✓

Irene has only given one piece of information so only gains 1 mark out of 2.

e feeds directly on plankton ✗

9 marks = Grade C answer

Grade booster ···> move a C to a B

Irene has made a number of simple careless mistakes. She has forgotten to label the graph and the biomass drawing. She also did not write down the name of the fish in part e. These simple mistakes have put her Grade C in doubt. You must keep concentrating in an examination and check all your answers.

GRADE A* ANSWER

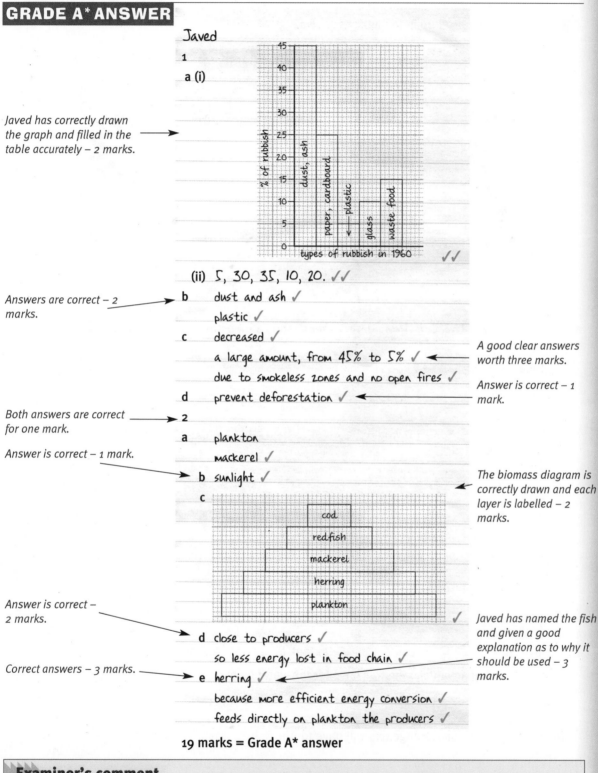

Javed

1

a (i)

Javed has correctly drawn the graph and filled in the table accurately – 2 marks.

(ii) 5, 30, 35, 10, 20. ✓✓

b   dust and ash ✓

Answers are correct – 2 marks.

plastic ✓

c   decreased ✓

a large amount, from 45% to 5% ✓ ←   *A good clear answers worth three marks.*

due to smokeless zones and no open fires ✓

d   prevent deforestation ✓ ←   *Answer is correct – 1 mark.*

2

Both answers are correct for one mark.

a   plankton

mackerel ✓

Answer is correct – 1 mark.

b   sunlight ✓   *The biomass diagram is correctly drawn and each layer is labelled – 2 marks.*

c

Answer is correct – 2 marks.

d   close to producers ✓

so less energy lost in food chain ✓   *Javed has named the fish and given a good explanation as to why it should be used – 3 marks.*

Correct answers – 3 marks.

e   herring ✓

because more efficient energy conversion ✓

feeds directly on plankton the producers ✓

19 marks = Grade A* answer

Examiner's comment

Javed has written excellent answers, not only showing good knowledge and understanding but also has avoided making simple errors.

Despite what some candidates think, it is possible to score maximum marks!

Living things in their environment

1 Four tubes were set up in the light as shown.

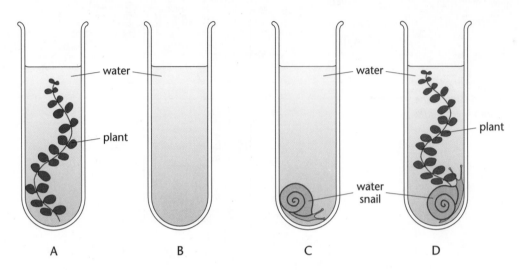

A   B   C   D

a) In which tube A, B, C or D would the water contain a high amount of carbon dioxide? Explain your choice.

Tube ....

Explanation ....

.... ③

b) In which tube A, B, C or D would the water contain a high amount of oxygen? Explain your choice.

Tube ....

Explanation ....

.... ③

c) Javed keeps goldfish in a water tank.
Irene tells him to put some water plants into the tank to help the fish.
Explain how this will help the fish.

....

....

.... ④

d) Javed says he should clean out the water tank about once a year.

i) Suggest what advantage this will have for the water plants.

....

.... ③

ii) Suggest what disadvantage this will have for the fish.

.... ①

TOTAL 14

2 The diagram shows how carbon compounds are recycled.

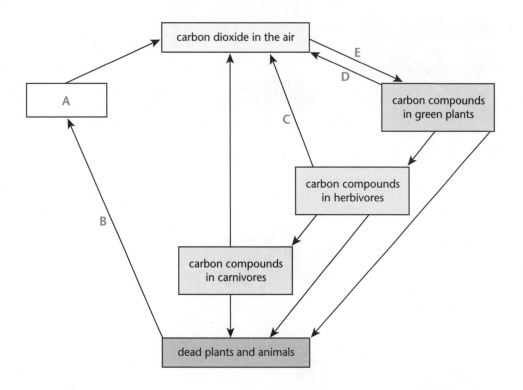

a) Name the processes shown by the arrows B, C, D and E.

B .... D ....

C .... E .... ④

b) Name a type of organism you expect to be named in box A.

.... ①

c) Explain why it is important for carbon to be recycled.

....

....

.... ③

d) State **one** other way in which the amount of carbon dioxide can be increased.

.... ①

e) Carbon dioxide is called a 'greenhouse gas' because it is involved in the greenhouse effect.
Explain what the greenhouse effect is and how it is caused.
One mark is awarded for a clear ordered answer and one mark is for the correct use
of scientific terms.

....

....

....

....

.... ⑤

TOTAL 14

Living things in their environment

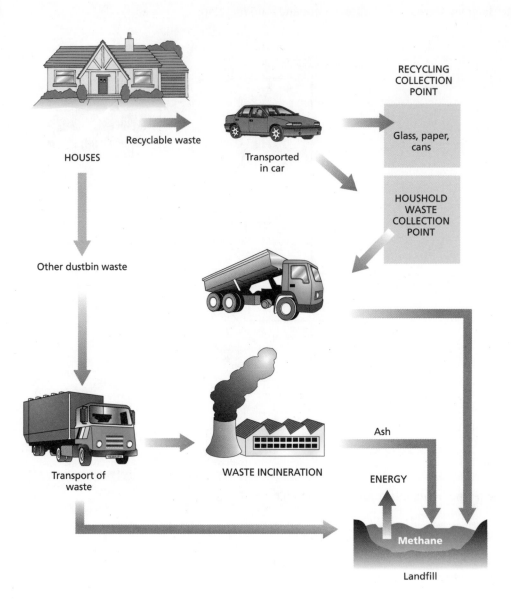

HOUSES

Recyclable waste

Transported in car

RECYCLING COLLECTION POINT

Glass, paper, cans

HOUSHOLD WASTE COLLECTION POINT

Other dustbin waste

Transport of waste

WASTE INCINERATION

Ash

ENERGY

Methane

Landfill

The diagram shows the disposal and recycling of household waste in a town.
The Town Council says,

'Not all waste is a problem'.

Discuss this statement.
Marks will be awarded for correct spelling, punctuation and grammar.

....
....
....
....
....
.... ⑥

TOTAL 6

4 The diagram shows two forms of the same species of moth.
One form is dark coloured and one is light coloured.

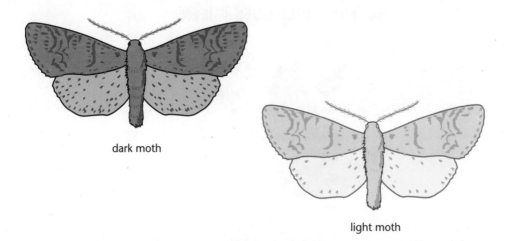

dark moth

light moth

Equal numbers of both forms were set free in a small park near the centre of a town.
The numbers of each form of moth caught in traps were counted each day.

Day	Number of dark moths caught	Number of light moths caught
1	20	20
2	20	18
3	18	10
4	12	6
5	10	2
6	10	0

a) Describe and explain the pattern of results for dark moths.

....

....

.... ③

b) Describe and explain the pattern of results for light moths.

....

....

.... ③

c) What results would you expect if equal numbers of the two forms of moth were released
in open countryside a long way from towns and cities?

....

....

.... ②

TOTAL 8

5 Read the following newspaper article.

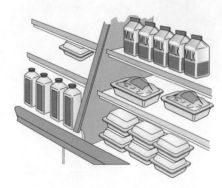

SUPERMARKET GOES GREEN.

A leading supermarket chain has announced a new biodegradable packaging. The new 'bio-packs' are made of starch. This easily breaks down, taking from 6 days to 6 weeks depending on conditions.

Environmental groups were initially pleased with the development but now say it will encourage more short-term litter in the streets since people will not bother to dispose of it properly. They also say that when it is put into landfill sites it will produce large amounts of gas.

a) What do you understand by the word 'biodegradable'?

....

.... ②

b) What conditions would cause the packaging to break down quickly?

.... ②

c) Suggest why this new packaging is better than:

  i) plastic bags

  ii) paper bags.

....

....

.... ③

d) Environmental groups are concerned that the new packaging could produce large amounts of gas.

  i) Name this gas.

.... ①

  ii) Why is this gas dangerous?

.... ①

  iii) How could this gas be useful?

....

.... ②

TOTAL 11

QUESTION BANK ANSWERS

① a) Tube C **①**
Snail respires
Producing carbon dioxide **②**
b) Tube A **①**
Plant photosynthesises
Producing oxygen **②**

EXAMINER'S TIP

With 3 marks available for each part a full explanation is required. You must name the two processes involved. Candidates find it difficult to understand the flow of gases in respiration and photosynthesis. Always remember that in sunlight, the exchange of gases in photosynthesis in plants is the reverse of respiration in animals.

c) Plants produce oxygen
By photosynthesis
Which the fish need
For respiration **④**
d) i) Waste products from fish
Will contain useful minerals, e.g. nitrogen
Which the plants will use for growth **③**
ii) May cause disease **①**

EXAMINER'S TIP

*These questions carried many marks, did you write down enough facts?
In part d), the question starts with 'Suggest...'.
This means different answers are possible so let your brain consider other areas of Biology to find an answer.*

② a) B is feeding/eating
or rotting/decay   D is respiration
C is respiration   E is photosynthesis **④**
b) Bacteria/fungi **①**

EXAMINER'S TIP

This question is frequently answered incorrectly. Candidates will either give a vague answer such as 'microscopic organisms' or name organisms such as scavenging animals.

c) Only small amount, 0.03% in air,
could all be used up by plants
And locked up in organisms for many years
(millions of years in coal) if not recycled **③**
d) Burning fossil fuels/peat/wood or
fermentation by yeast to produce alcohol **①**
e) Earth's atmosphere acts like a greenhouse
So Earth is warmed up (global warming)
Light energy (short-wavelength) gets through
Earth's atmosphere
Warm objects on Earth give out heat energy
(long wavelength/infra red radiation)
Which is trapped
3 marks for any three points
2 extra marks for using good English and
correct scientific terms **⑤**

EXAMINER'S TIP

Where extra marks are available for the quality of written expression, you must carefully plan your answer to make sure your English is good (sentences, logical sequence) as well as using the correct scientific words such as 'global warming', 'infra red radiation'.

③ Waste as a problem ...
Transport causing noise, pollution, traffic congestion
Landfill using up land, destroying scenery, using agricultural land
Using up energy to burn waste, fuel for lorries
2 good points for two marks **②**
Waste being useful...
Recycling of glass, paper, aluminium
Methane gas used as a fuel **②**
extra two marks awarded for writing in at least two sentences, using capital letters and full stops. Examiners usually allow only one incorrect spelling **②**

EXAMINER'S TIP

In discussing a situation, you must write about both sides to present a balanced view. In this case you are not asked to decide if it is right or wrong so there is no need to express an opinion. Since 6 marks are on offer it is very important to plan your answer by jotting down some ideas before you start writing your answer, otherwise it could easily become a rambling vague account worth few marks.

Living things in their environment

④ a) Declined/fewer/from 20 to 10
      Levelled out after 5 days
      Some eaten   ❸
   b) Declined/fewer/from 20 to 0
      None then caught/no moths left
      All eaten   ❸
   c) Reverse results
      Dark moths not camouflaged, white moths
      are camouflaged   ❷

EXAMINER'S TIP

This is a well known piece of research to show Natural Selection at work. In dark industrial surroundings the dark moths will be camouflaged and few will be seen and eaten by birds. The paler moths will easily be seen and will be eaten by birds. The fewer the number of moths the fewer caught.

If the information is not known to you, do not panic! Read the information carefully, think about your answers and you can get full marks.

⑤ a) Broken down
      Naturally/by bacteria/fungi   ❷
   b) Warm, wet, plenty of oxygen
                                   any two points   ❷
   c) Better than plastic because it quickly
      breaks down
      And plastic takes many years
      Better than paper because it avoids
      deforestation   ❸
   d) i) Methane   ❶
      ii) Flammable/catch alight   ❶
      iii) Collected and burnt
          To provide heat energy   ❷

EXAMINER'S TIP

This type of question is often put in examination papers. Sometimes the information is real, sometimes it is made up! In this case it is real, the supermarket chain being Tesco. With unfamiliar situations, always remember to keep cool and do not panic. Read the article at least twice to make sure you understand it. The questions are then usually straightforward.

CHAPTER 6

Adaptations

To revise this topic more thoroughly, see Chapters 7 and 12 in *Letts Revise GCSE Biology Study Guide*.

 Try these GCSE questions and then compare your answers with the Grade C and Grade A model answers on page 57.

1 The diagrams show two types of fish.

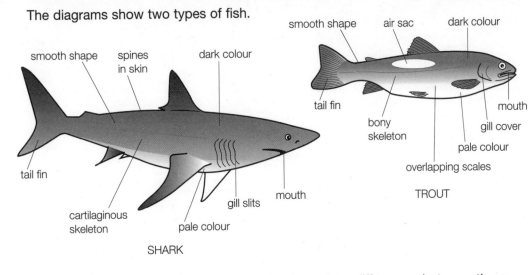

SHARK

TROUT

a Using the information in the diagram, write down **four** differences between these two types of fish.

....

....

....

.... **[4]**

b (i) Using the information in the above diagrams, write down **three** features, found in both fish, which are adaptations to life in water.

....

....

....

.... **[3]**

(ii) Write down **one** other adaptation, not shown in the above diagrams, to life in water.

.... **[1]**

2 The basic pattern of a vertebrate limb is called a pentadactyl limb.
   The diagram shows the limb bones of five different animals.

a Write on the lines the name of each animal.
   Choose from:

   bat   bee   bird   human   lizard   whale   worm

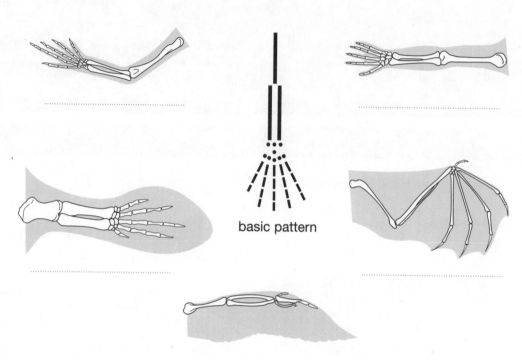

basic pattern

[5]

b Choose **one** of these limbs and describe how it is different from the basic pattern.
   Explain how the changes help the animal.

   ....

   ....

   .... [2]

3 Complete the following information by drawing lines to connect up the type of
   tissue with its property and position. One line has been drawn for you.

Property	Tissue	Position
absorbs shock not rigid	ligament	joins bone to bone
very strong some flexibility	cartilage	joins muscle to bone
strong a little flexibility	tendon	on ends of bone

[3]

(Total 18 marks)

These two answers are at Grades C and A. Compare which one your answer is closest to and think how you could have improved it.

GRADE C ANSWER

Kevin

Kevin has written two good answers but then has forgotten to write a comparison answer for the tail in the trout and the skeleton in the shark. Two marks out of four.

1

a shark, mouth underneath trout, mouth at front ✓ shark, gill slits trout, gill cover ✓ shark has unequal tail, trout has a bony skeleton ✗

A good answer – 3 marks.

b (i) smooth streamlined shape ✓ gills ✓

Kevin has written down two good answers but three were required.

(ii) large number of eggs produced to ensure survival ✓

Kevin has mixed up human and lizard bones. The lizard has very long claws, humans do not.

2

a human ✗ lizard ✗ whale ✓ bat ✓ bird ✓

b it has very long fingers to form a wing ✗

Unfortunately Kevin has not written down which animal limb (bat) he is describing so no marks can be awarded.

Kevin has shown he does not know much about cartilage, tendons and ligaments. Candidates are often confused with these names. Only 1 mark.

3 absorbs shock → cartilage
very strong → ligament ✓
ligament → joins muscle to bone ✗
tendon → joins bone to bone ✗
cartilage → on ends of bone ✓

10 marks = Grade C answer

Grade booster ⟶ move a C to a B
When asked for a comparison, you must write down information about both things.
When asked to select an organism, you must declare which one you are writing about.

GRADE A ANSWER

Lindsay

1

a shark no air sac, trout has air sac ✓
shark spines in skin, trout overlapping scales ✓
shark cartilaginous skeleton, trout bony skeleton ✓

Lindsay's first three answers are correct. Her answer about size of fins is too vague. Three marks out of four.

shark small fins, trout large fins ✗

b (i) streamlined shape ✓ gills ✓
colour camouflage ✓

Answers are all correct – 3 marks.

(ii) lateral line to detect movement ✓

2

Answers are all correct. Bees and worms do not have skeletons – 5 marks.

a lizard ✓ human ✓ whale ✓ bat ✓ bird ✓

b whale, short thick bones ✓
to form a large flipper to swim with ✓

Answers are all correct – 3 marks.

Other answers are Bird, fewer/longer fingers to form a large wing to fly Lizard, long claws to dig/grasp/climb.

3 absorbs shock → cartilage very strong →
ligament ✓ ligament → joins bone to bone ✓
cartilage → on ends of bone ✓
tendon → joins muscle to bone

17 marks = Grade A answer

Examiner's comment
Lindsay's answers are excellent with a near maximum score. Well done.

1. Lindsay has an accident at school.
She is taken to hospital for an X-ray on her leg.

a) i) Label parts A and B.

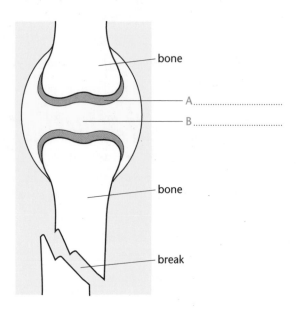

bone

A...

B...

bone

break

②

ii) Describe the functions of parts A and B.

....

.... ②

b) Bones break only in extreme situations.
Complete the following sentences about bones.
Choose from:

bones	**calcium phosphate**	**dead**	**joint**	**lever**	**living**	**ligaments**	**starch**

Bones are strengthened by deposits of ....

Broken bones repair themselves because they contain .... cells.

Sprains happen when .... are torn.

Dislocations happen when a bone is forced out of a .... ④

c) Your backbone is made up of 34 vertebrae.
Suggest why there are so many.

....

....

.... ②

TOTAL 10

2 Plant leaves lose water through small holes called stomata.
Gases also enter and leave through these holes.
Stomata are usually in the lower surface of leaves.
The diagrams show cross-sections through various leaves showing how they are adapted to different conditions.

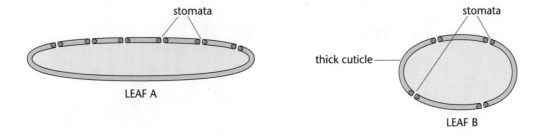

LEAF A

stomata

stomata

thick cuticle

LEAF B

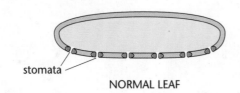

stomata

NORMAL LEAF

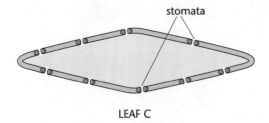

stomata

LEAF C

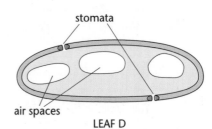

stomata

air spaces

LEAF D

a) i) Using A, B, C or D match up the following types of leaves.

   Upright grass leaf ....

   Floating leaf ....

   Submerged leaf ....

   Desert leaf ....   ④

  ii) Select **one** leaf, apart from the normal leaf, and explain the importance of its adaptation.

   ....

   ....

   ....   ②

b) Many trees are adapted to survive the winter by getting rid of their leaves.
   Explain why.

   ....

   ....   ②

TOTAL 8

Adaptations

3   Mammoths are in the news.

Mammoths are extinct. Despite this a lot of information is known about mammoths because some have been found frozen in glaciers.

Mammoths created vast grasslands, eating 90 kg of plants each day.

Scientists are uncertain why mammoths became extinct.

3 million years ago
Mammoths evolve

100 000 years ago
Modern man evolves

15 000 years ago
Ice Age ends

3700 years ago
Mammoths extinct

Today
Frozen mammoths
discovered in Russia

5 m

a)   In terms of temperature regulation, what advantage to the mammoth was there in being very big?

....

....

.... ③

b)   Suggest **two** other reasons why mammoths were so successful.

....

....

.... ②

c)   i)   How long ago did mammoths become extinct?

.... ①

    ii)   Suggest possible reasons why mammoths became extinct.

....

....

.... ③

TOTAL 9

4 The diagram shows the life cycle of a mosquito.

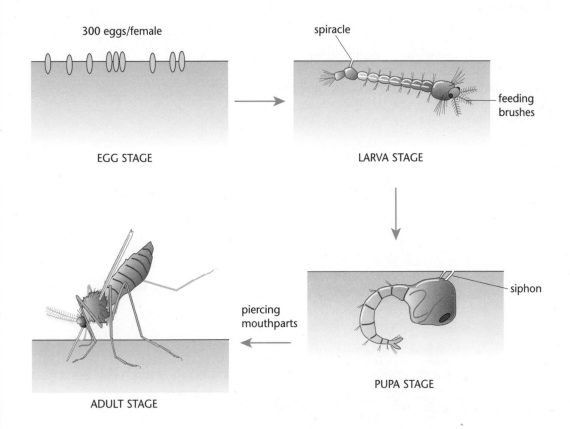

300 eggs/female

EGG STAGE

spiracle

feeding brushes

LARVA STAGE

siphon

PUPA STAGE

piercing mouthparts

ADULT STAGE

a) How is the egg stage and larva stage adapted to life in water?

Egg ....

Larva .... ②

b) Suggest why it is an advantage for different stages of the mosquito to live in different environments.

....

....

.... ②

c) The adult mosquito transmits the disease malaria.

i) What part of the human body is affected by malaria and what are the results?

....

....

.... ②

ii) The adult mosquito can be controlled by spraying with insecticide.
Suggest how the other stages may be controlled.

....

....

.... ②

TOTAL 8

Adaptations

5 Normally plants die when there is very little water.

a) Explain why.

....

....

.... ③

b) Cactus plants are adapted to live in dry conditions.
The diagram shows four special adaptations.

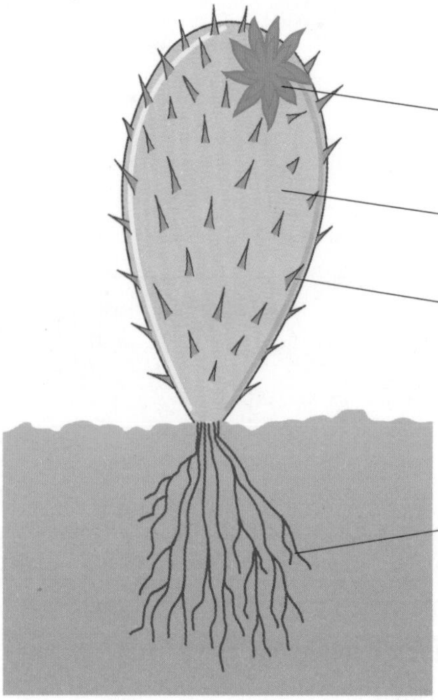

flowers produce large numbers of seeds

swollen stem

leaves reduced to spines

deep root system

Explain how each adaptation helps the cactus plant.

Flowers ....

Stem...

Leaves ....

Roots .... ④

c) Deserts can be very hot during the day but very cold at night.
Suggest why small burrowing mammals are able to survive these conditions.

....

....

.... ③

d) The kidney controls water loss from animals.
Suggest why the kidney tubules are very long in desert animals.

....

....

.... ③

TOTAL 13

1 a) i) A is cartilage
B is synovial fluid   **2**

ii) A protects ends of bone/stops bones
rubbing together
B cushions/absorbs shock/lubricates
the joint   **2**

b) calcium phosphate
living cells
ligaments
joint   **4**

EXAMINER'S TIP

These questions are straightforward recall of information. If you have revised thoroughly you should get full marks. However, don't be too confident. Work carefully through the questions.

c) For flexibility/many joints
Specialisation/different types of vertebrae
have different functions   **2**

EXAMINER'S TIP

Did this question cause you problems? It is certainly not a straightforward recall of information, the word 'suggest' tells you that. You may have to leave this type of question for a few minutes and then return to it. Your brain may then come up with a possible answer.

2 a) i) Leaf C
Leaf A
Leaf D
Leaf B   **4**

ii) Leaf C, stomata on both surfaces, leaf upright so no advantage of stomata on lower surface
Leaf A, stomata only on upper surface, they would be waterlogged on the lower surface
Leaf D, few stomata, with air spaces since leaf underwater. Flow of water through plant is not a problem
Leaf B, few stomata, thick cuticle so little water is lost in transpiration. Little water available in desert

only one answer required   **2**

EXAMINER'S TIP

Did you remember to name the leaf you were describing? If not, you do not score any marks. Try to avoid this silly mistake.

b) Leaves thin with little protection
Easily damaged by frost/snow, so disease enters tree   **2**

EXAMINER'S TIP

This apparently simple question may have caused you problems! Try to answer questions like this by turning them around, e.g. 'What would happen to leaves in winter conditions?', and then you think of the answer.

3 a) Large volume compared to surface area
So heat energy conserved
So mammoths could survive very cold conditions   **3**

b) Herbivores/plenty of food
Thick fur to protect from cold
Large tusks to dig out plants or to protect themselves
Few enemies
No major diseases   any two reasons   **2**

EXAMINER'S TIP

The surface area/volume ratio is important in many areas of Biology so make sure you understand it.
Any 'success' question can always be answered in the same format. Has it enough food? Are there few enemies to eat it? Are there any major diseases?

c) i) 3700 years ago   **1**
ii) Temperature change
Killed by human hunters
Vegetation changed
Wiped out by disease   any three points   **3**

EXAMINER'S TIP

Any 'extinction' question can also be answered in the same format. Are there changes in temperature, gases, water? Does its food run out? Does a different predator appear? Does a disease wipe them out?

Adaptations

④ a) Egg, many eggs produced/floats on surface
Larva, spiracle to get air/feeding brushes to
filter food **②**

b) They will not compete with each other
For space/food/oxygen **②**

b) Flowers, large number of seeds to ensure
some germinate
Stem, swollen with stored water
Leaves, reduced to spines to avoid water
loss in transpiration
Roots, deep system to reach more water **④**

EXAMINER'S TIP

*Part b) is an interesting question for a Biologist.
The evolution of a larval stage may occur
separately from the adult stage, leading to
many possibilities. However, you should have
realised that competition was the answer.*

EXAMINER'S TIP

*Did you realise that you needed good answers?
The question would be far too easy if answers
like 'swollen stem' and 'small leaves' were
accepted as full answers.*

c) i) Blood **①**
Red blood cells destroyed
Lack of oxygen
High fever/death   any one point **①**

ii) Draining swamps/pools/ponds/ditches
Introducing an animal e.g. fish to eat them
Spraying surface of water with oil so they
cannot respire   any two points **②**

c) Small, do not overheat/lose heat quickly
Burrowing, avoids extreme heat and cold
Mammal, constant body temperature **③**

d) Reabsorb
More water
To produce concentrated urine so little
water lost **③**

⑤ a) Water needed to keep cells turgid/their
shape
For transport of food/hormones/waste
For chemical reactions to occur in solution
Required for photosynthesis
     any three points **③**

64

Adaptations

CHAPTER 7

Food and microbes

To revise this topic thoroughly, see Chapter 8 in *Letts Revise GCSE Biology Study Guide.*

Try these sample GCSE questions and then compare your answers with the Grade C and Grade A model answers on page 67.

1   The graph shows the change in temperature of fish between unloading from trawlers and their sale in shops.

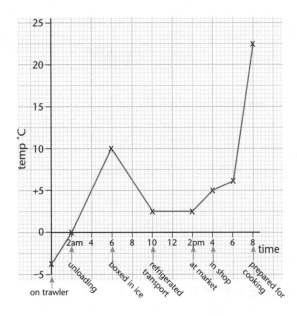

a   How long did it take from catching fish to preparing it for cooking?

.... [1]

b   At which stage were the fish the most safe from bacterial growth?
Explain your answer.

....

....

.... [3]

c (i) At which stage were the fish most at risk from bacterial growth?
Explain your answer.

....

....

.... [3]

(ii) Suggest what you could do to avoid this risk.

.... [1]

2   A famous scientist, Louis Pasteur, carried out research on wine and milk.
He wanted to find out what caused them to go bad.
He used soup in special flasks.

When cooled the soup did not go bad

FLASK A

Soup boiled to
make it sterile

When the flask's neck was
removed the soup went bad

FLASK B

a   What is meant by the word 'sterile'?

.... [1]

b   Explain why the soup in flask A did not go bad.

....

....

.... [3]

c   Explain why the soup in flask B did go bad.

....

.... [3]

d   Suggest and explain what would happen to the soup in flask A if it had not been
boiled but heated to either 25°C or 75°C.

At 25°C ....

.... [3]

at 75°C ....

.... [2]

(Total 20 marks)

These two answers are at grades C and A. Compare which one your answer is closest to and think how you could have improved it.

GRADE C ANSWER

Mohammed

Mohammed did not include the units (hours) so no mark is awarded.

Again Mohammed has correctly identified the stage as being prepared for cooking, but has not explained the effect of temperature on bacterial growth.

A good answer – 1 mark.

A poor answer since 3 marks are available. Mohammed did not explain that bacteria would grow in the soup.

1
a 20 ✗
b on trawler ✓   because frozen ✓
c (i) being prepared for cooking ✓
     because temperature is highest ✓
  (ii) keep refrigerated until just before
       cooking ✓
2
a bacteria cannot grow ✓
b bacteria ✓
   could not move through S bend ✓
c open to the air ✓
d At 25°C soup goes bad ✓
   At 75°C soup stays fresh for some
   time ✓

Mohammed has correctly identified the stage as being on the trawler, but has not fully explained why the fish are safe. 2 out 3 marks.

Mohammed has not realised that the bacteria would first be killed by boiling. 2 marks out of 3.

The answers are correct but too brief. Only 1 mark 3 for each.

11 marks = Grade C answer

Grade booster ···> move a C to a B
Mohammed has made a common mistake in not writing in the units. Mohammed has a fairly good understanding about bacteria but has written very brief answers to many questions.

GRADE A ANSWER

Noreen

Noreen has made a mistake in reading the information from the graph. She has simply looked at the number 8 on the time axis. No marks.

A very good answer – 3 marks.

1
a 8 hours ✗
b on trawler ✓   because frozen ✓
   bacteria cannot grow ✓
c (i) being prepared for cooking ✓ high temperature of
      22°C ✓ encourages bacterial growth ✓
  (ii) prepare it quickly, keeping it cold ✓
2
a no bacteria ✓
b bacteria ✓   could not enter through S bend ✓
   those present at first are killed by boiling ✓
c bacteria ✓ could enter from the air ✓
   produce waste products in soup ✓
d at 25°C soup bad ✓ quickly ✓
   rapid bacterial growth ✓ at 75°C soup
   fresh ✓ for a short time because some bacteria
   killed ✓

A very good answer – 3 marks.

Noreen has realised she must explain what happens in the soup to get the third mark.

Good answers.

19 marks = Grade A answer

Grade booster ···> move A to A*
Noreen has an excellent understanding of bacterial action. A pity she made a simple error in reading information from the graph, otherwise she would have scored maximum marks.

Food and microbes

67

Food and microbes

1 A scientific laboratory tests food samples.

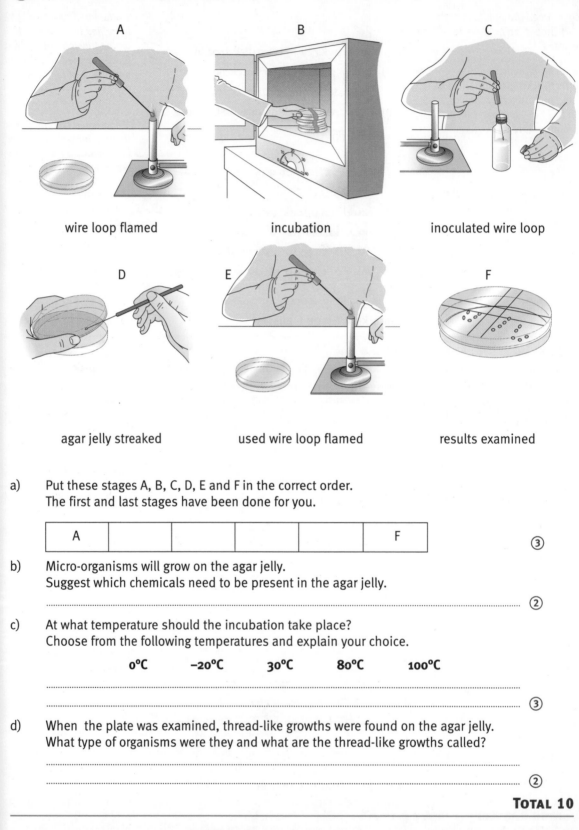

A

wire loop flamed

B

incubation

C

inoculated wire loop

D

agar jelly streaked

E

used wire loop flamed

F

results examined

a) Put these stages A, B, C, D, E and F in the correct order.
The first and last stages have been done for you.

A					F

③

b) Micro-organisms will grow on the agar jelly.
Suggest which chemicals need to be present in the agar jelly.

....

②

c) At what temperature should the incubation take place?
Choose from the following temperatures and explain your choice.

      0°C   –20°C   30°C   80°C   100°C

....

....

③

d) When the plate was examined, thread-like growths were found on the agar jelly.
What type of organisms were they and what are the thread-like growths called?

....

....

②

TOTAL 10

2 Food can be preserved by many different methods.

a) Draw lines to connect up each method with its action and food example.

Action	Method	Food
Heated then sugar added	freezing	corned beef
Cooled to −18°C	canning	fish fingers
Heated to 121°C and sealed	drying	vegetables
Placed in hot air	irradiation	coffee
Exposed to gamma radiation	bottling	fruit

⑧

b) Name **one** other method of food preservation and explain how it works.

....

.... ②

TOTAL 10

---

3 A company was growing yeast to sell.
They investigated the effect of altering the amount of oxygen on yeast growth.
The investigation was carried out at 20°C in a 5 dm³ flask containing a sugar solution.

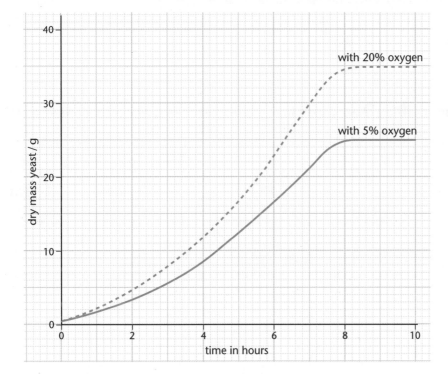

a) i) What was the final yeast yield with:

5% oxygen ....

20% oxygen .... ?

②

ii) Explain why the yeast needed oxygen.
Marks will be awarded for the correct use of technical terms.

....

....

....

....

....

.... ⑤

b) Which yeast grew faster?
Explain your answer.

....

.... ②

c) i) For how long were the yeasts growing?

.... ①

ii) Suggest reasons why the yeasts stopped growing.

....

.... ②

d) If the yeast was supplied with no oxygen, it would not die immediately.

i) Explain why not.

.... ①

ii) Name the chemicals which the yeast would produce.

.... ②

e) Name **two** commercial uses of yeast.

1 ....

2 .... ②

f) Suggest and explain a reason, apart from external conditions, why yeast grows so quickly.

....

....

.... ②

TOTAL 19

4 Various microbes can be grown in large vessels called fermenters.

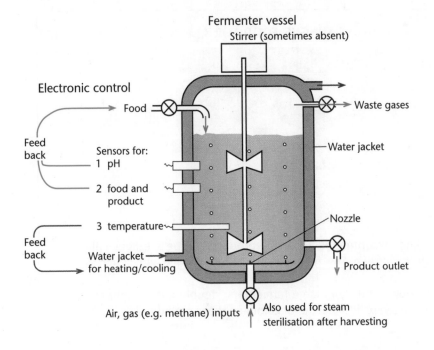

Fermenter vessel

a) Explain the importance of the feedback mechanisms for pH and food.

....

....

....

.... ④

b) If the temperature is too high the microbes are killed.

Explain why.

....

....

.... ③

c) Explain an advantage of using a stirrer.

....

.... ②

d) Suggest why the word 'fermenter' may be the wrong word to describe the vessel used.

....

.... ②

e) Explain why strict sterilisation stages are necessary in this process.

....

....

.... ③

TOTAL 14

5 Read the following article.

FUNGUS FOOD FOR ALL

Proteins called mycoproteins are being produced using a fungus called *Fusarium graminearum*. The fungus needs to be provided with starch and ammonia.

The fungus grows quickly in large fermenters, doubling its weight every five hours. It has very little flavour or colour so these are added artificially.

---

a) Fungi are organisms which do not make their own food. Describe how this differs from green plants.

....

....

....

....

.... ④

b) Suggest why the fungus needs ammonia.

....

.... ②

c) The fermenter contains 10 kg of fungus.
How much fungus will there be after 15 hours?
Show your working.

....

....

.... ③

d) Describe **one** advantage and **one** disadvantage of using mycoprotein.

....

....

.... ②

TOTAL 11

❶ a) C, D, E, B   ❸

EXAMINER'S TIP

The standard way of marking a sequence question if the answer is partly incorrect is to look at the relationship between two letters. Therefore if part of your answer is incorrect, it is marked: C before D = 1, D before E = 1, E before B = 1. So an answer C, E, D, B would score 2 marks. Always double check the sequence to make sure it works.

b) Food
    Named food, e.g. sugar, glucose, starch, protein or vitamins/minerals   ❷

c) 30°C
    Enzymes work best at this temperature
    Too high or too low will denature enzymes   ❸

EXAMINER'S TIP

The last point will only be realised by few candidates. It is easy to forget that there are three marks for this question, one for selecting the temperature and two for an explanation. Remember that enzymes are denatured NOT killed.

d) Fungi/fungus
    Hyphae/mycelium   ❷

❷ a) Heated to bottling
    Cooled to freezing
    Heated 121°C to canning
    Placed to drying   ❹
    Exposed to irradiation   1 right = 1,
        2 right = 2, 3 right = 3, 4 or 5 right = 4
    Freezing to fish fingers
    Canning to corned beef
    Drying to vegetables/accept fruit
    Irradiation to coffee
    Bottling to fruit/accept vegetables   ❹

b) Chemical preservation using sulphur dioxide/vinegar which kills microbes
    Or UHT which uses steam to kill microbes and their spores   ❷

EXAMINER'S TIP

There are a lot of marks (8) available, so take your time and select carefully. Use a pencil to first connect up the words in case you make a mistake or realise you are wrong when the last answer does not fit. Then remember to replace them with straight lines in ink. It is amazing the number of candidates who take their pen for a walk around the paper!

❸ a) i) 25 g
    35 g
    if units are missing deduct one mark   ❷

EXAMINER'S TIP

Numbers in an answer must always include the units, in this case g. In an answer like this, the candidate is only penalised once.

  ii) Uses it to break down
    Sugars/food
    Releasing energy
    Technical terms such as oxidise, aerobic respiration   ❺

EXAMINER'S TIP

An equation summing up respiration would have been acceptable.
sugar + oxygen = carbon dioxide + water + energy

b) i) Uses anaerobic respiration   ❶
  ii) Alcohol
    Carbon dioxide   ❷

c) Yeast with 20% oxygen
    More oxygen releases more energy   ❷

d) i) 8 hours (Did you remember to include the units?)   ❶
  ii) Temperature of 20°C too low
    All sugar/food used up
    Build up of waste products
    Becomes too acidic/pH changes
            any two points   ❷

e) Brewing beer/wine
    Bread making   ❷

f) Uses asexual reproduction/budding
    Very quick method of reproduction   ❷

a) Controls levels/amounts
   Explanation of feedback, e.g. too much so less supplied, too little more supplied
   Different pH will affect enzymes, work best at specific pH levels
   Levels of food will affect production of yeast, i.e shortage of food will cause less to be made ❹

b) Enzymes
   Denatured (**not killed!**)
   Metabolism slows down/stops or membranes destroyed ❸
c) Avoids hot spots in temperature
   Which would kill microbes
   Or, ensures good contact
   With microbe and food/oxygen ❷
d) Fermenter implies use of fermentation
   Which does not use oxygen ❷
e) Kill unwanted microbes
   Which would compete for food/oxygen
   And produce different products ❸

a) Photosynthesis
   Green chloroplasts
   Using light, carbon dioxide and water
   Producing sugars/starch ❹

b) Provide nitrogen
   For amino acids/protein ❷
c) After 5 hours = $10 \times 2 = 20\,kg$, after 10 hours
   $= 20 \times 2 = 40\,kg$, after 15 hours
   $= 40 \times 2 = 80\,kg$
   Answer $= 80\,kg$
       2 marks for answer 1 mark for unit ❸

d) Advantage: very quick production/does not need light/can be made to look like different foods
   Disadvantage: little flavour/colour ❷

CHAPTER 8

Microbes and disease

To revise this topic more thoroughly, see Chapter 10 in *Letts Revise GCSE Biology Study Guide.*

 Try these sample questions and then compare your answers with the Grade C and Grade A model answers on page 77.

1 An unknown disease has affected a school. Pupils suffering from the disease are coughing, have chest pains, a high temperature and a skin rash. Not all pupils are suffering from the disease.

The graph shows the body temperature of a pupil with the disease.

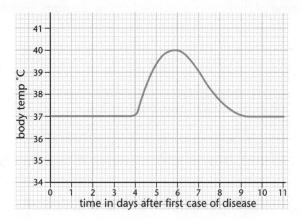

a How long did it take for this pupil to show a symptom of the disease?

.... **[1]**

b How long did this symptom last?

.... **[1]**

c Suggest an explanation why some pupils did not catch the disease.

.... **[2]**

d (i) Suggest a likely method of transmission of this disease.

.... **[2]**

(ii) Name **two** other ways diseases can be acquired. Give an example of each way.

1 ....

2 .... **[4]**

2   The diagram shows the body's response to an invading pathogen.

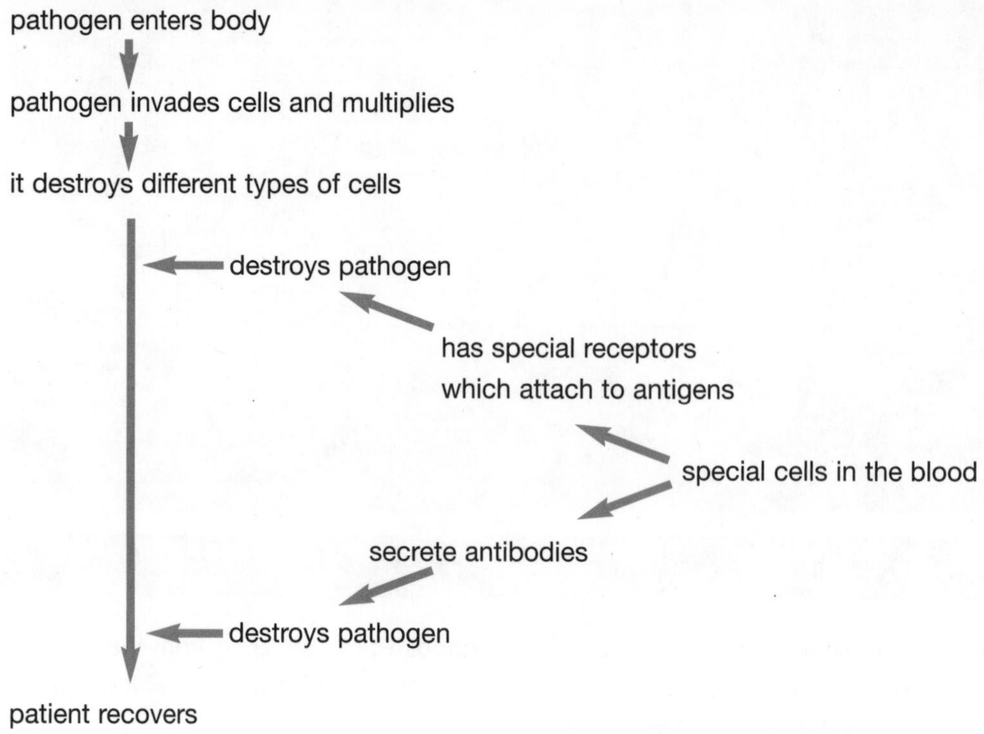

pathogen enters body

↓

pathogen invades cells and multiplies

↓

it destroys different types of cells

← destroys pathogen

has special receptors
which attach to antigens

special cells in the blood

secrete antibodies

← destroys pathogen

patient recovers

a   Suggest the type of pathogen described in the diagram. Explain your choice.

.... [2]

b (i) Name the special cells in the blood involved in this response.

.... [1]

(ii) Where else in the body would these cells be found?

.... [1]

(iii) Name the two different types of cells which are produced by the special cells
in the blood.

1..   2 .... [2]

c   Explain the importance of memory cells in the response to an infection.

....

....

.... [3]

d   Old people are offered vaccination against influenza.

Suggest why a different vaccine is given to these people each year.

....

.... [3]

(Total 22 marks)

These two answers are at Grades C and A. Compare which one your answer is closest to and then think how you could have improved it.

GRADE C ANSWER

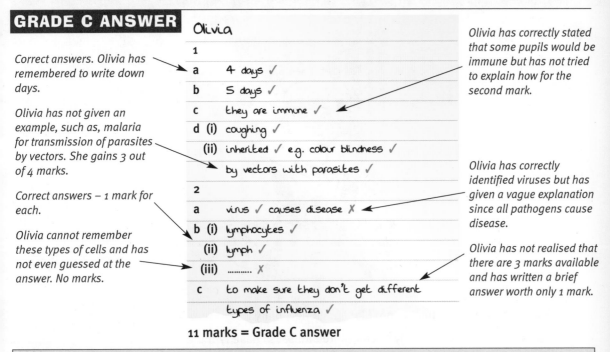

Olivia

1

a   4 days ✓

b   5 days ✓

c   they are immune ✓

d (i) coughing ✓

  (ii) inherited ✓ e.g. colour blindness ✓

    by vectors with parasites ✓

2

a   virus ✓ causes disease ✗

b (i) lymphocytes ✓

  (ii) lymph ✓

  (iii) .... ✗

c   to make sure they don't get different

    types of influenza ✓

11 marks = Grade C answer

Correct answers. Olivia has remembered to write down days.

Olivia has not given an example, such as, malaria for transmission of parasites by vectors. She gains 3 out of 4 marks.

Correct answers – 1 mark for each.

Olivia cannot remember these types of cells and has not even guessed at the answer. No marks.

Olivia has correctly stated that some pupils would be immune but has not tried to explain how for the second mark.

Olivia has correctly identified viruses but has given a vague explanation since all pathogens cause disease.

Olivia has not realised that there are 3 marks available and has written a brief answer worth only 1 mark.

Grade booster ····⟩ move a C to a B

Olivia was probably a little careless in not writing down an example in d(ii). Question **2** is more demanding. She should have tried a guess for the names of the cells, she might have been lucky! She followed this with a weak answer, possibly being upset by not knowing the answer for the previous question.

GRADE A ANSWER

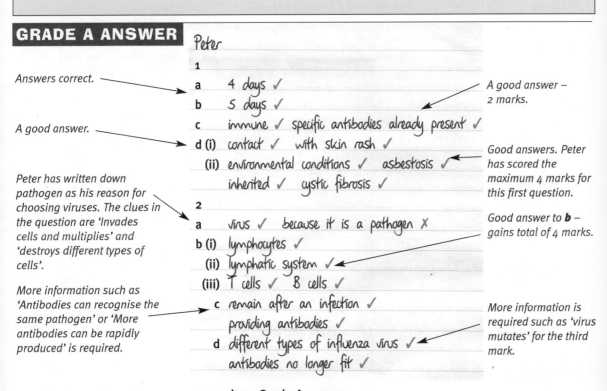

Peter

1

a   4 days ✓

b   5 days ✓

c   immune ✓ specific antibodies already present ✓

d (i) contact ✓   with skin rash ✓

  (ii) environmental conditions ✓   asbestosis ✓

    inherited ✓   cystic fibrosis ✓

2

a   virus ✓   because it is a pathogen ✗

b (i) lymphocytes ✓

  (ii) lymphatic system ✓

  (iii) T cells ✓   B cells ✓

c   remain after an infection ✓

    providing antibodies ✓

d   different types of influenza virus ✓

    antibodies no longer fit ✓

19 marks = Grade A answer

Answers correct.

A good answer.

Peter has written down pathogen as his reason for choosing viruses. The clues in the question are 'Invades cells and multiplies' and 'destroys different types of cells'.

More information such as 'Antibodies can recognise the same pathogen' or 'More antibodies can be rapidly produced' is required.

A good answer – 2 marks.

Good answers. Peter has scored the maximum 4 marks for this first question.

*Good answer to **b** – gains total of 4 marks.*

More information is required such as 'virus mutates' for the third mark.

Grade booster ····⟩ move A to A*

Peter has written some excellent answers to Question **2**. He thoroughly understands the body's immune reactions. Well done! More information in **2 c** and **d** would have increased his marks.

1

Read the following extracts from Mary Mallone's diary.

March	1895	My doctor in Ireland says I have typhoid.
June	1900	I emigrate to America and get a job as a cook in a large house.
August	1900	One of the family in the house becomes ill with typhoid.
December	1900	I am sacked but soon get another job as a cook.
July	1901	One of my kitchen assistants dies of typhoid.
April	1902	I get another job as a cook.
June	1902	Eleven of the family get typhoid.
January	1905	I have had many jobs as a cook but get sacked when people become ill.
August	1907	Dr Soper believes I am a typhoid carrier. Tests prove it. He says that my hands become contaminated when I go to the toilet, transferring the typhoid bacteria.
February	1908	I am being kept in hospital.
April	1910	I am released from hospital.
April	1915	I have a job as a cook in a New York hospital but a typhoid outbreak occurs so I am sacked. I am being kept in hospital and not allowed out.

Mary Mallone died in 1938.

a) What type of organism causes typhoid?

.... ①

b) How was the disease transmitted?

....
.... ②

c) How many people caught typhoid from Mary and died?

.... ①

d) What is a carrier of disease?

....
....
.... ③

e) Mary was kept in hospital for many years even though she did not need treatment.

If she had been released in 1915 what two precautions should she have taken to avoid passing on the disease?

....

.... ②

TOTAL 9

2 A cholera epidemic occurred in the southern part of London in 1854.

Water was supplied by two water companies, who got water from the River Thames.

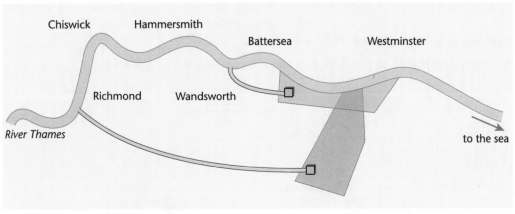

Key

██ area supplied by
Southwark Water Company

██ area supplied by
Lambeth Water Company

Water company	Number of houses supplied	Number of deaths from cholera	Death rate / 10 000 houses
Southwark	40 000	286	
Lambeth	26 107	14	5.4

a) Calculate the death rate/10 000 houses for the Southwark Water Company and fill in the empty box in the table.
Show your working.

....

.... ②

b) i) Which company provided the safer water?
Explain your answer.

.... ①

ii) Suggest how the water supplied by the two water companies was different.

....

....

.... ③

c)   What simple method could have been used to make the water safe to drink in 1854?

.... ①

d)   Name **two** reasons why our tap water is now safe to drink.

1 ....

2 .... ②

e)   Cholera often occurs after an earthquake.
      Suggest why.

      ....

      .... ②

TOTAL 11

---

3   Agar jelly in a Petri dish was left for a few hours.

A lid was then placed on the Petri dish and it was incubated for a few days.

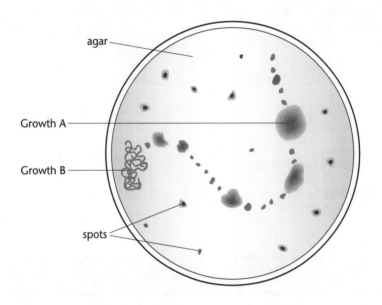

a)   What types of organisms are likely to be in:

      growth A ....

      growth B .... ? ②

b)   Suggest how they could have colonised the agar jelly.

      ....

      .... ③

c)   Suggest how other growths, shown as spots on the agar jelly, could have got there.

      .... ①

d)   What precautions should be taken when handling the Petri dish?

      ....

      .... ②

TOTAL 8

QUESTION BANK ANSWERS

① a) A bacterium **❶**
b) Not washing hands after going to the toilet then handling food **❷**
c) One **❶**

EXAMINER'S TIP

Never be surprised by an unusual question such as a diary of Mary Mallone. Read the information carefully and slowly. Using a ruler to force you to read one line at a time can help you to avoid missing important information. Did you write down 13 for part c), counting the people who were infected with typhoid rather than those who died?

d) Has the disease
Can pass it on
But does not suffer from it/does not show symptoms **❸**
e) Not being a cook/not handling food
Always washing hands after visiting toilet **❷**

EXAMINER'S TIP

The key word in this question is 'suggest', which means a variety of answers are acceptable. However, the answers must be feasible. Many candidates write down impossible answers such as, 'Never go to the toilet', showing they have not given much thought to the question!

② a) $\frac{286}{4} = 71.5$ **❷**

EXAMINER'S TIP

This is a difficult question to answer. Many candidates would have divided by 40 000 instead of by 4. One line has been done for you so an answer of 0.007 15 would look wrong. Once you have calculated an answer always check to see if it is within reason.
As in all calculations, show your working and use a calculator.

b) i) Lambeth Water Company
Fewer deaths/lower death rate **❶**

EXAMINER'S TIP

Where there are only two choices, no mark is given for the answer. The correct selection is a lead up to an explanation. This is done to try to avoid you forgetting to name which company you are writing about.

ii) Water taken from different places
Lambeth took water from higher up the R. Thames
Less polluted/contaminated
By sewage/industrial waste
any three points **❸**
c) Boiled
(Heated not accepted as an answer) **❶**
d) Water chlorinated
Sewage treated before putting back into river **❷**
(An answer containing reference to anti-pollution laws is acceptable)
e) Water contaminated
By sewage pipes/water pipes being broken **❷**

EXAMINER'S TIP

These questions require a lot of thought to work out possible answers. You are being required to apply your knowledge.
'Thinking time' is built into the length of an examination paper, simply writing down the answers would take only about half the allocated time. So do not panic if you do not immediately know the answer, there is a little time to think about it.

③ a) Bacteria
Fungi/fungus **❷**
b) Animal/insect
Carrying micro-organisms/spores
Walking across agar **❸**
c) From the air **❶**

EXAMINER'S TIP

Since the trail is not random the large growths could only have come from something (an animal) leaving a contaminated trail across the agar.
The random spots are therefore from exposure to the air.

d) Dish and lid should be taped together
Never opened
Disposed of safely
Warning signs/notices displayed
any two points **❷**

FOR MORE INFORMATION ON THIS TOPIC ... SEE REVISE GCSE BIOLOGY ... CHAPTER 10

Microbes and disease

CHAPTER 9

Food production

To revise this topic more thoroughly, see Chapter 13 in *Letts Revise GCSE Biology Study Guide.*

Try this sample GCSE question and then compare your answers with the Grade C and Grade A model answers on page 84.

1   The charts show the types of cultivation and their levels of production in a small country.

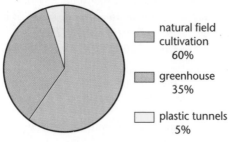

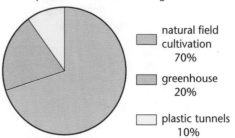

a   Use the information in the first chart to complete this table.

Type of cultivation	% cultivation	Area of cultivation in hectares
natural field		
greenhouse	35	1050
plastic tunnel		

Use this space for working out.

[4]

b Look at both charts.

(i) Which type of cultivation is the **most** productive/hectare?
Explain your choice.

....

.... **(2)**

(ii) Which type of cultivation is the **least** productive/hectare?
Explain your choice.

....

.... **(2)**

c Explain how climate can affect crop production.
Marks will be given for the quality of written communication.

....

....

....

....

.... **(6)**

2 The following information shows the range of pH that different crops grow well in.

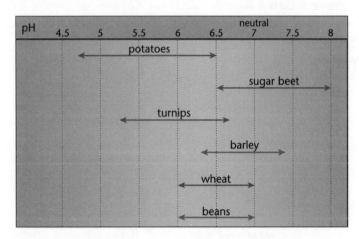

a (i) Which crop will grow best in an alkaline soil?

.... **(1)**

(ii) Which crop will grow best in an acidic soil?

.... **(1)**

b (i) Suggest how soil can become acidic.

.... **(1)**

(ii) How can this acidity be neutralised?

.... **(1)**

(Total 18 marks)

These two answers are at grades C and A. Compare which one your answer is closest to and think how you could have improved it.

GRADE C ANSWER

Quereshi has correctly realised the % cultivation figures but has used the wrong figures to work out the percentages. He gains 2 marks out of 4.

Quereshi has written a good answer, gaining two extra marks. However, he has only written about one aspect of climate. 4 marks out of 6.

Quereshi

1

a   natural field 60 ✓   50 ✗
    plastic tunnel 5 ✓   600 ✗
    working; 3000/60 = 50,
    3000/5 = 600 ✗

b (i) natural field ✗
       because it produces more ✗

  (ii) plastic tunnel ✗
        because it produces least ✗

c   a warm temperature of 25°C ✓
    will increase photosynthesis and enzyme
    activity and their growth
    and crop yield ✓✓✓

2

a (i) sugar beet ✓   (ii) potatoes ✓
b (i) a lot of peat in the soil ✓
  (ii) adding lime ✓

*Quereshi has not understood the question and has taken the highest and lowest crop production figures.
He has not realised that 5% of land for plastic tunnels gives 10% of production. No marks.*

Answers to 2 all correct – 4 marks.

10 marks = Grade C answer

Grade booster ⋯⋯> move a C to a B
Quereshi has shown a weakness in mathematics and data handling. Candidates are expected to be able to carry out calculations involving simple mathematics.

GRADE A* ANSWER

Answers to 1 all correct and working shown – 4 marks.

A well written answer so two extra marks are awarded. Maximum marks gained, i.e. 6 marks.

Reshma

1

a   natural field 60 ✓   1800 ✓
    plastic tunnels 5 ✓   150 ✓
    60/100 × 3000 = 1800
    35/100 × 3000 = 150

b (i) plastic tunnels ✓
       from 5% land get 10% production ✓
  (ii) greenhouse ✓
        from 35% land get only 20% production ✓

c   high winds, flooding, frost and snow ✓✓
    can destroy crops ✓ desert conditions such as high
    day temperatures and very low night temperatures
    will result in poor crop levels ✓

2

a (i) sugar beet ✓   (ii) turnips ✓
b (i) too much peat ✓   (ii) adding lime ✓

Good answers, scores total of 4 marks.

Answers all correct – 4 marks.

18 marks = Grade A* answer

Examiner's comment
Reshma has written excellent answers showing not only a good grasp of factors affecting plant growth but also good mathematical skills.

1 Plants are grown in containers lacking in various minerals.

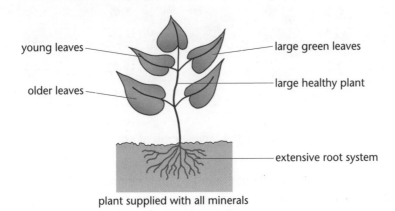

young leaves — large green leaves
older leaves — large healthy plant
— extensive root system

plant supplied with all minerals

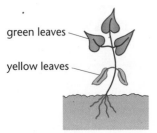

green leaves
yellow leaves

plant without nitrates

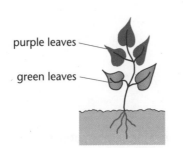

purple leaves
green leaves

plant without phosphates

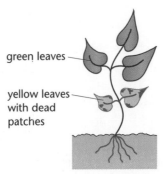

green leaves
yellow leaves with dead patches

plant without potassium

a) Describe the effects of the lack of nitrates, phosphates and potassium on plant leaves.

....

....

....

.... ⑥

b) Suggest why the plants lacking in nitrates and phosphates are smaller than normal.

....

.... ②

c) Outline how chemicals such as nitrates, phosphates and potassium become available to plants.

....

....

....

.... ④

TOTAL 12

Food production

2 Two crops, potatoes and wheat, were grown in two halves of a large field.
Different parts of the field were given different amounts of fertilisers.

Treatment	Fertiliser in kg/hectare			Crop yield in tonnes/hectare	
	nitrates	phosphates	potassium	potatoes	wheat
A	0	0	0	8.47	1.69
B	96	0	0	8.30	3.68
C	0	77	107	16.63	2.04
D	96	77	107	38.57	6.60

a) Which treatment A, B, C or D produced the largest crop of both potatoes and wheat?

.... ①

b) The percentage increase in crop yield of potatoes with treatment D was 355.4%
Calculate the % increase for wheat.
Show your working.

....

....

.... ②

c) Describe the pattern of results for potatoes when different amounts of fertilisers were used.

....

....

.... ③

d) The design of the experiment has been criticised by some scientists.
How can it be improved?

....

....

.... ③

e) i) Adding too much fertiliser to fields can cause problems.

Explain what these problems could be.

....

....

....

.... ③

ii) If natural fertiliser such as manure is used, fewer problems are caused. Explain why.

....

.... ②

TOTAL 14

3 Early man used a 'slash and burn' method of agriculture.
Small areas of forest were cleared to grow crops.
After a few years of growing the same crops the crop yield fell.
The families then moved on and cleared another area of forest.

a) Suggest how burning the plants helped to produce good crops for the first few years.

....

....

.... ③

b) Explain why the crop yields fell after a few years.

....

.... ②

c) Later farmers used crop rotation so they could stay in the same place.

 i) Explain what crop rotation is and how it works.

....

....

.... ③

 ii) A leguminous plant called clover is often used in crop rotation.
 At the end of the year it is ploughed back into the ground.
 Explain why.

....

....

.... ③

TOTAL 11

Food production

❶ a) Nitrates – smaller leaves
older leaves yellow
Phosphates – younger leaves purple
older leaves green but smaller
Potassium – older leaves normal/large and
green
older leaves yellow with dead patches **❻**

EXAMINER'S TIP

You were asked to describe the leaves so do not waste time in describing the roots or stem, there are no marks allocated for these comments.

b) Nitrates and phosphates used to make
proteins/enzymes/DNA
For growth **❷**
c) Plants and animals die
Tissues mainly protein/muscle
Decay/rot
By bacteria/fungi
Dissolve in water
Fertilisers   any four points **❹**

❷ a) D **❶**
b) Increased by 6.60 – 1.69 = 4.91
4.91/1.69 × 100 = 290.5% **❷**

EXAMINER'S TIP

This is a difficult calculation, since many candidates would think their answer should be 100 or less. A calculator is necessary for this question, the answer has been given correct to 2 dp. As in all calculations, a correct answer, even without working out, would be awarded full marks.

c) When nitrates were added, yield decreased
When phosphates and potassium added, yield
increased/doubled
When all three minerals added, yield
increased/by over four times **❸**
d) Fixed levels of fertilisers have been used
Different levels/amounts should have been
investigated
Yields when only potassium is added missing
Yields when only phosphates are added
missing
Other conditions in the field, such as pH,
water levels, exposure to wind/sun should
have been checked to make sure it was a 'fair
test'   any three points **❸**

EXAMINER'S TIP

This question is examining 'Ideas and Evidence in Science'. Your experience in course work would have been useful in realising that a number of other variables should have been investigated.

e) i) Leeching out/washing out
Into rivers/lakes
Causing eutrophication
Rapid growth of plants/algae
Rot and use up oxygen
Death of animals in water/fish
Affecting drinking water
       any three points **❸**
ii) Rotting/decaying
Much slower/continual release **❷**

❸ a) Minerals
In ash
Named minerals, e.g. nitrates, phosphates,
potassium
An alternative answer could be
Heat
Kills pests/controls disease
So more plants survive **❸**

EXAMINER'S TIP

This is another question examining 'Ideas and Evidence in Science'. In a question like this it is a good idea to read all the questions first since they will help you to get involved in the story as well. Question b) helps you to work out a possible answer to question a).

b) Minerals used up
By same crop
or
Pests and diseases
Built up in the soil because of same crop **❷**
c) i) Changing crops in one area/field
Usually a three or four year rotation
Because different plants use different levels of
different minerals
Also avoids build up of specific pests/diseases
       any three points **❸**
ii) Clover's roots
Contain nodules full of bacteria
Which are nitrogen fixing
They convert nitrogen in air to nitrates/
increase nitrogen content
       any three points **❸**

EXAMINER'S TIP

Questions c) i) and ii) are another example of how reading all the questions before answering any question can be useful. After reading both questions, you will probably realise that the nitrogen cycle is involved.

CHAPTER 10

Genetics and genetic engineering

To revise this topic more thoroughly, see Chapter 11 in *Letts Revise GCSE Biology Study Guide*.

Try these sample GCSE questions and then compare your answers with the Grade C and Grade A model answers on page 91.

Two scientists, James Watson and Francis Crick, worked out the chemical structure of deoxyribonucleic acid (DNA). They described the structure as being like a ladder.

The sides are made of sugar and phosphate groups and the rungs of four different bases.

The bases are adenine, thymine, cytosine and guanine.

a   Label the diagram of DNA.

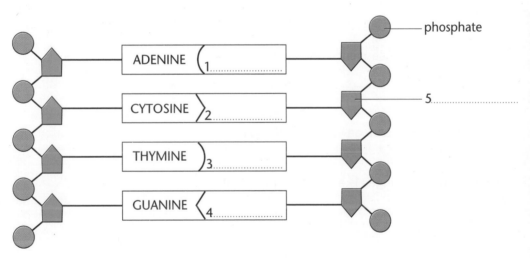

[5]

b   What are the two jobs of DNA?

1 ....

2 .... [2]

c Cells use another chemical called ribonucleic acid (RNA).

State **two** differences between DNA and RNA.

....

....

....

.... [2]

d Cells use RNA to make proteins from amino acids.

(i) How many bases are used in the code for one amino acid?

.... [1]

(ii) In which cell structure are proteins made?

.... [1]

(iii) Why is the production of proteins so important?

....

....

.... [2]

e This is a sequence of bases.

ATC   GTA   CCG   ATA   TAC

It has changed to:

ATC   CCG   ATA   TAC   GTA

(i) How has the sequence changed?

....

.... [2]

(ii) What is this change in sequence called?

.... [1]

(iii) Why are changes like these so important?

....

....

.... [2]

(Total 18 marks)

GRADE C ANSWER

Susan

Susan has not realised that there is a fifth label to identify.

a   1 thymine ✓ 2 guanine ✓ 3 adenine ✓
    4 cytosine ✓

b   carries genetic information to pass on
    to next generation ✓

Susan has written a good answer but has only described one job of DNA.

Susan has not identified which one has two or one strand so she is not given the first mark.

c   one has two strands the other only one
    strand ✗
    thymine is replaced by uracil in RNA ✓

Susan has wrongly identified mitochondria as being where proteins are made.

d (i) three ✓
  (ii) mitochondria ✗

She has realised that proteins are important to make enzymes but has not written enough for the second mark.

  (iii) enzymes are proteins ✓
        many different enzymes ✗

These answers are all correct.

e (i) GTA sequence ✓
        has been moved/translocated ✓
  (ii) mutation ✓
  (iii) disrupts the whole gene ✓

A better explanation is required for the second mark.

12 marks= Grade C answer

Grade booster ···▷ move a C to A B

Susan has shown very good understanding of this topic but she has made some simple errors. She did not double check her answers or she would have realised that she had missed out one answer in **a**. In writing a comparison, candidates frequently make the same mistake as Susan by not identifying which one is being compared. By checking the number of marks available for questions, Susan could well have achieved a Grade A.

GRADE A ANSWER

Trevor

All correct.

a   1 thymine ✓   2 guanine ✓   3 adenine ✓
    4 cytosine ✓   5 sugar ✓

All correct.

b   carry coding information ✓   to copy itself ✓

A good answer for two marks.

c   DNA has two strands in a helix, RNA has
    only one strand ✓ In RNA thymine is
    replaced by uracil ✓

d (i) three ✓
  (ii) ribosomes ✓

Trevor has not fully explained how enzymes are important, e.g. speeding up chemical reactions, so scores only 1 mark.

  (iii) to make enzymes ✓

e (i) GTA ✓ has been translocated into a different
        position ✓

Trevor has the importance of mutations in a different way from Susan.

  (ii) mutation ✓
  (iii) to produce variation ✓
        for natural selection and evolution to occur ✓

Good answers.

17 marks = Grade A answer

Examiner's comment

Trevor has written excellent answers. He has obviously learnt a lot of information and read the Grade Booster comments on previous pages!

Genetics and genetic engineering

1 *Craterostigma plantigineum* is called a 'Resurrection Plant' because it can recover from long periods of drought.

Its genes control enzymes which change a sugar called octulose into a different sugar called sucrose.

In very dry conditions the sucrose forms crystals preventing the cells from being damaged.

Scientists are now trying to make other plants such as cereals behave in the same way.

a) Explain how this could be useful.

....

.... ②

b) Scientists are planning to use genetic engineering techniques to make cereals behave like the resurrection plant.

Outline the stages required.

One mark is awarded for a clear ordered answer and one mark is for the correct use of scientific terms.

....

....

....

....

....

.... ⑥

c) Many people are concerned about the use of genetic engineering.

i) Suggest why.

....

....

.... ③

ii) Write down **two** arguments supporting the use of genetic engineering.

....

.... ②

TOTAL 13

2 Cloning can be used as a type of reproduction.

a) Explain how it differs from sexual reproduction.

....

.... ②

b) The following statements describe cloning.

    A DNA is extracted.

    B DNA from another individual is added.

    C An ovum is selected.

    D The ovum is returned to female individual.

The correct order of these statements is:

③

c) Apart from being able to make copies of yourself, describe **one** situation where cloning could be useful.

....

.... ②

TOTAL 7

3 The diagram shows RNA in action.

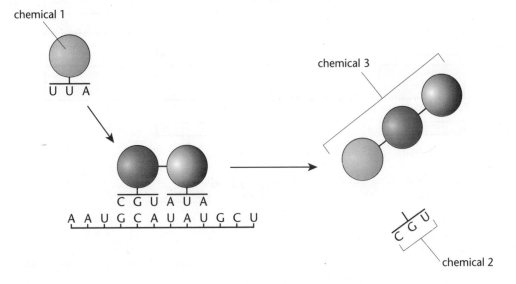

a) In which part of the cell does this take place?
Choose from:

        cytoplasm   **membrane**   **mitochondria**   **nucleus**

.... ①

b) Name the chemicals 1, 2 and 3.

Chemical 1 ....

Chemical 2 ....

Chemical 3 .... ③

c) Explain what process is shown by the diagram.

....

....

....

.... ④

TOTAL 8

④ Police are investigating the theft of examination papers.
They find a half eaten apple at the scene of the crime.
They carry out a 'DNA fingerprint' test on saliva found on the apple.
They take DNA samples from three people believed to have carried out the theft.

Band	DNA result from crime scene	DNA result from suspect 1	DNA result from suspect 2	DNA result from suspect 3
1		▭		▭
2	▭		▭	▭
3	▭	▭	▭	▭
4		▭		▭
5	▭		▭	▭
6	▭	▭	▭	▭
7	▭		▭	

a) Suggest why DNA samples can be used to identify people.

....

.... ②

b) Which band is common to all four results?

.... ①

c) Which suspect do you think carried out the theft?
Explain your answer.

....

.... ②

TOTAL 5

1 a) So crops/food
Can be grown in dry/desert conditions /can survive drought ❷
b) Gene identified
Removed from DNA
By enzymes
DNA of new plant/cereal plant
Opened up by enzymes
Gene inserted
Enzymes repair DNA
Any four points plus two marks if answer is in sequence and well written in sentences ❻

EXAMINER'S TIP

It is important to write down these stages in the correct sequence since there is a total of six marks available, almost a grade difference. Jot down in the margin as many facts as you can remember, then organise them in the correct sequence. You must also write your answer in good English, including complete sentences.

c)(i) Unknown risks/outcome cannot be known
Could create new diseases
Against God/nature ❸
(ii) Cures for human diseases
Advantages in food production/new flavours/ better protection against pests/disease ❷

EXAMINER'S TIP

This type of question differentiates between a Grade C and A candidate. The weaker candidate tends to write only one fact for each side of the argument.
In such an important topic as this you must revise thoroughly.

2 a) In cloning there is no change in chromosome number, while in sexual reproduction the chromosome number is halved in the gametes
In cloning there is no chromosome breaking and recombination as in meiosis in sexual reproduction
In cloning the new individuals are identical to the original, following sexual reproduction the new individuals are unique
any two points ❷

EXAMINER'S TIP

In a comparison question you must write down answers which show a comparison. If you write, 'In cloning the new individuals are identical', you must also write down what happens in sexual reproduction, otherwise you will not be awarded the mark. This is a very common mistake.

b) C A B D ❸

EXAMINER'S TIP

These sequencing questions can be difficult to answer correctly, so always write your answer in pencil first, then check the whole sequence before writing it in pen/biro.
Examiners usually mark these questions by looking for the position of two letters at a time, C before A = 1 mark, A before B = 1 mark, B before D = 1 mark.

c) Saving endangered species
e.g. Giant Panda so they do not become extinct
or
Recreating extinct species
e.g. Dodo or Mammoth to restore gene diversity ❷

3 a) Cytoplasm ❶
b) 1 Amino acid
2 RNA
3 Polypeptide/protein ❸
c) Making proteins
From coding in DNA/RNA
tRNA takes an amino acid to ribosome
3 letter code matches up with correct part of mRNA
Amino acids link up making protein
any four points ❹

EXAMINER'S TIP

This question is set at a Grade A level and demands a good understanding of the topic. Many candidates simply write down a vague description of the diagram without realising the point of the process, i.e. to make proteins/enzymes.

Genetics and genetic engineering

4 a) Very large number/billions of possible
   combinations
   So individuals are unique ❷
b) Band 6 ❶
c) Suspect 2
   Bands match so it is his saliva found at the
   crime scene ❷

FOR MORE INFORMATION ON THIS TOPIC ... SEE REVISE GCSE BIOLOGY ... CHAPTER 11

Centre number	
Candidate number	
Surname and initials	

𝓛𝓮𝓽𝓽𝓼 Examining Group
General Certificate of Secondary Education

Biology
Paper 1

Higher tier

Time: one hour

Instructions to candidates

Write your name, centre number and candidate number in the boxes at the top of this page.

Answer ALL questions in the spaces provided on the question paper.

Show all stages in any calculations and state the units. You may use a calculator.

Include diagrams in your answers where this may be helpful.

Use a sharp pencil when drawing a graph or diagram.

Allocate a few minutes towards the end of the examination to check your answers.

Information for candidates

The number of marks available is given in brackets [2] at the end of each question or part question.

The marks allocated and the spaces provided for your answers are a good indication of the length of answer required.

In certain questions extra marks are available for the quality of your written answer.

1   Amanda looks at a sample of water with a microscope.
    She sees an organism called Euglena.

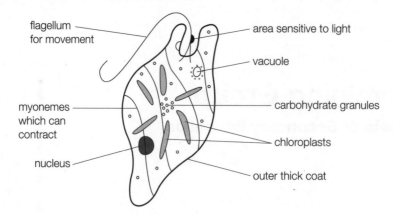

She does not know whether it is a plant or an animal.

(a) Write down **two** features of Euglena which are usually found in animals.

1 ....

2 .... **[2]**

(b) Write down **one** feature of Euglena which is only found in plants.

   .... **[1]**

(c) Write down **one** feature of Euglena which is found in both plants and animals.

   .... **[1]**

(d) Amanda puts a few drops of water containing Euglena into three dishes
    containing water.
    The dishes are kept in different conditions.

Condition	dish 1	dish 2	dish 3
Light	✓	✓	✗
Carbon dioxide	0.5%	0.0%	0.5%
Temperature	20°C	20°C	2°C

Samples were examined with a microscope.

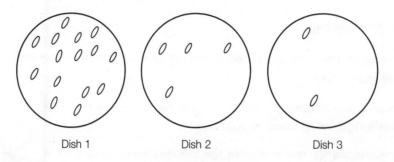

Dish 1   Dish 2   Dish 3

(i) Count the number of Euglena in each sample and write the results in the table.

Dish	Number of Euglena
1	
2	
3	

[2]

(ii) Explain these results.
One mark is awarded for a clear ordered answer.

....

....

....

....

....

.... **[4]**

(iii) Amanda's friends did not believe her results.
To convince them, how could the experiment be improved?

....

....

....

.... **[2]**

(iv) Amanda thinks she will get similar results if microscopic animals are used.
Explain why she will be wrong.

....

....

.... **[2]**

(Total 14 marks)

[turn over

2 Benjamin goes to the doctors.
The doctor ties a special cuff around his arm to measure his blood pressure.
Benjamin sees small bulges appear in some of his blood vessels.

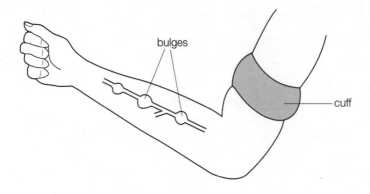

bulges

cuff

(a) (i) What type of blood vessel will show these bulges?

.... [1]

(ii) How are these bulges formed?

....

.... [2]

(b) The doctor measures Benjamin's blood pressure.

(i) How is blood put under pressure?

....

.... [1]

(ii) What type of blood vessel will contain blood under high pressure?

.... [1]

(iii) Suggest why blood needs to be under high pressure in these blood
vessels.

....

....

.... [2]

(c) Some babies are born with a 'hole in the heart'.
Their heart can look like this.

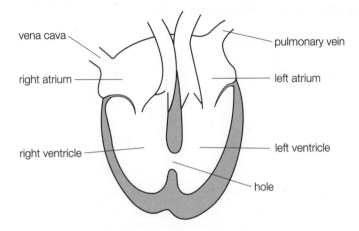

(i) What difference is there between the blood in the vena cava and the blood in the pulmonary vein?

....

.... **[2]**

(ii) Explain what problems the 'hole in the heart' will cause to the baby.

....

....

.... **[3]**

(iii) Serious heart defects may require a heart transplant.
People are encouraged to carry a donor card so their organs can be donated when they die.
Some people think this should be made compulsory.

Discuss the advantages and disadvantages of making the carrying of donor cards compulsory.

One mark is awarded for a clear ordered answer and one mark is for the correct use of scientific terms.

....

....

....

....

....

.... **[6]**

(Total 18 marks)

[turn over

3   As food travels through the body it passes through many organs.

(a) Put the following organs in the correct sequence.

The first one has been done for you.

A   oesophagus (gullet)

B   large intestine

C   mouth

D   stomach

E   small intestine

C				

[3]

(b) Enzymes act on food.
Draw a line to connect up each food type with the enzyme which acts on it and what is then produced.

Enzyme	Food type	What is produced
amylase	starch	fatty acids and glycerol
lipase	protein	glucose
protease	fat	amino acids

[4]

(c) Bile is produced in the liver.
The diagram shows the action of bile on fats.

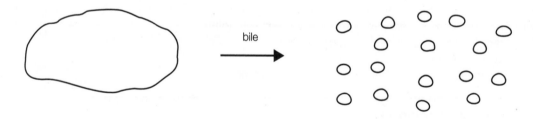

large drop of fat   small droplets of fat

(i) What is this action called?

.... [1]

(ii) Why is this change useful?

....

.... [1]

(Total 9 marks)

Letts

4 The figures in the table show the population numbers of Kaibab deer.

Time	1900	1910	1920	1930	1940	1950
Population in thousands	20	24	40	25	20	18

(a) Plot this information on the grid.
Draw a 'line of best fit'.

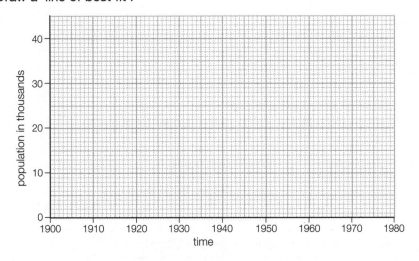

[3]

(b) In 1910 a large scale programme to kill off mountain lions and wolves was started in the same area.

(i) Suggest links between mountain lions, wolves and Kaibab deer.

....

.... [1]

(ii) What happened to the Kaibab deer population between 1920 and 1950? Explain this change.

....

....

.... [2]

(iii) Using your graph, predict how the population of Kaibab deer will change from 1950 to 1980.
Explain your answer.

....

....

.... [2]

(Total 8 marks)

[turn over

5   Luke puts his hand on a nail.

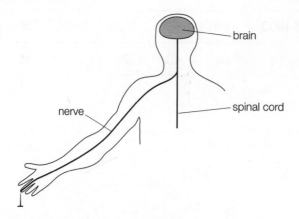

He quickly moves his hand and shouts in pain.
The table contains information on the timing of the sequence of events.

Time in milliseconds	What took place
0	Luke puts his hand on a nail
30	Impulse reaches spinal cord
43	Luke moves his hand
80	Impulse reaches brain
110	Luke shouts in pain

(a) How long did it take for Luke to:

move his hand .... milliseconds

shout in pain .... milliseconds.   **[1]**

(b) Suggest why there is a difference in these times.

....

....   **[2]**

(c) What type of nervous reaction is involved when Luke moves his hand?

....   **[1]**

Letts

(d) Complete the following sentences.

Choose from,

brain   **electrical**   **effector**   **motor**   **physical**

receptor   **sensory**   **spinal cord**

When Luke puts his hand on a nail, pressure is put on a ....…............. .

Nerve impulses then travel along a …...................... neurone.

These impulses are …..........…............ .

These impulses will go first to the …...........…......... .   **[4]**

(e) After a while, Luke's face goes pale with shock.
This reaction is due to hormones.

(i) What are hormones?

....

....   **[1]**

(ii) Explain why this reaction took a few minutes.

....

....

....   **[2]**

(iii) Name **one** other hormone, apart from sex hormones, which is produced
by the body.
Explain its function.

Name of hormone …...................…........

Function ....

....

....   **[2]**

(Total 13 marks)

[turn over

6 Using a microscope Anita looks at the lower surface of two leaves.
They had been kept in different conditions.

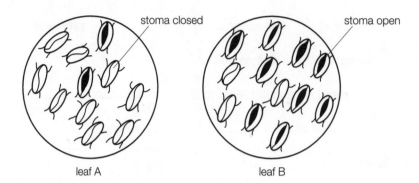

leaf A   leaf B

(a) Complete the table to show the number of open and closed stomata in each leaf.

	Leaf A	Leaf B
Number of open stomata		
Number of closed stomata		

[2]

(b) Describe the importance of stomata to plants.
One mark is awarded for a clear ordered answer.

....

....

....

....

....

....

.... **[5]**

(c) Which leaf will have been kept in dry dark conditions?
Explain your answer.

Leaf ....

....

....

.... **[2]**

(d) Anita grows some plants from seeds.

When the seedlings have grown about 10 cm she transplants them into large pots.

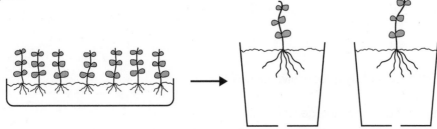

(i) Suggest why the plants will grow better in the large pots.

....

....

.... **[3]**

(ii) She puts a polythene bag over each plant and pot for a few days.

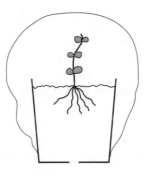

Suggest how this helps the plant.

....

....

.... **[3]**

(Total 15 marks)

[turn over

7 Marfan's syndrome is a rare inherited condition.
People who have Marfan's syndrome have very long arms, legs, fingers and toes.

The condition is carried by a recessive allele (r).

(a) Two parents both have the genotype Rr.

   (i) What will be their phenotype?

   .... **[1]**

   (ii) Complete the genetic cross for these parents.

	parent	
	R	r
parent		

parent

                                                                                 [3]

   (iii) What genotype will result in a person with Marfan's syndrome?

   .... **[1]**

   (iv) Scientists have found that there are more people with Marfan's syndrome than was expected.
Suggest why.

   ....

   .... **[1]**

(b) Marfan's syndrome also affects ligaments in the eye.

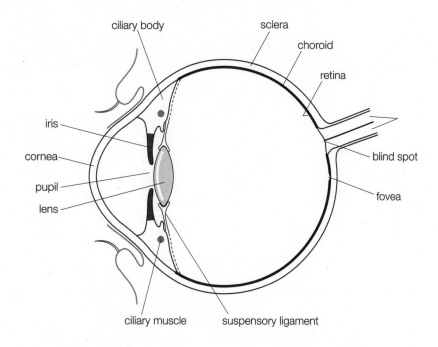

(i) Explain how light from an object reaches the retina.

....

....

.... **[3]**

(ii) How does information leave the eye?

....

....

.... **[2]**

(iii) If the ligaments and ciliary muscles do not work, how will sight be affected?

....

....

.... **[2]**

(Total 13 marks)

(Overall total = 90 marks)

Letts Examining Group

General Certificate of Secondary Education

Biology
Paper 2

Higher tier

Time: one and a half hours

Instructions to candidates

Write your name, centre number and candidate number in the boxes at the top of this page.

Answer ALL questions in the spaces provided on the question paper.

Show all stages in any calculations and state the units. You may use a calculator.

Include diagrams in your answers where this may be helpful.

Use a sharp pencil when drawing a graph or diagram.

Allocate a few minutes towards the end of the examination to check your answers.

Information for candidates

The number of marks available is given in brackets **[2]** at the end of each question or part question.

The marks allocated and the spaces provided for your answers are a good indication of the length of answer required.

In certain questions extra marks are available for the quality of your written answer.

Letts

EDUCATIONAL

1 Penguins are flightless birds.
They are adapted to live in very cold conditions

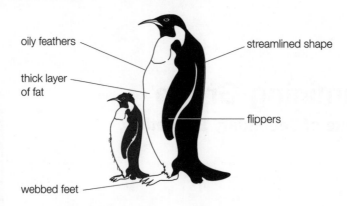

oily feathers

thick layer
of fat

webbed feet

streamlined shape

flippers

(a) Select **two** of the labelled features and explain how each helps the penguin to survive the cold.

Feature 1 ....

Explanation ....

....

Feature 2 ....

Explanation ....

.... **[4]**

(b) Penguins produce only one chick each season.
Discuss the advantages and disadvantages of this.
One mark is awarded for a clear ordered answer and one mark is for the correct use of scientific terms.

....

....

....

....

....

.... **[4]**

(c) The bones of birds are hollow.
Explain why.

....

.... **[2]**

(d) The limb bones of five different animals are shown below.

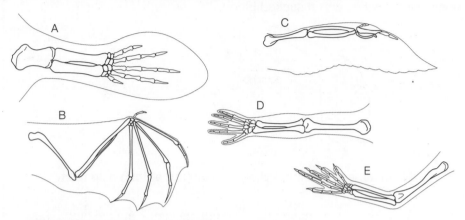

Which set of limb bones belongs to a bird?
Explain your choice.

Limb bone set ....

Explanation ....

.... **[2]**

(e) How penguins became flightless can be explained using Darwin's ideas on evolution.

   (i) The following sentences explain Darwin's views.
   Complete the boxes, putting the sentences in the correct order.

   A those offspring with beneficial variations survive and breed

   B the large number of offspring show many small variations

   C all species reproduce to excess

   D those offspring with beneficial variations become the majority of the
   population

C			

   [2]

   (ii) Lamarck, another scientist, had a different explanation.
   The following sentences explain Lamarck's views.
   Complete the boxes putting the sentences in the correct order.

   A the new feature is passed on to the next generation

   B species in a population are very similar

   C some offspring develop a beneficial feature in their lifetime

   [2]

   (Total 16 marks)

   [turn over

2 Hundreds of years ago, doctors removed blood from sick people.
This was done using leeches which sucked blood.

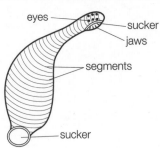

(a) Explain why this treatment was more likely to harm rather than help them.

....

.... **[2]**

(b) In more recent times, doctors tried to replace blood using direct transfusions from one person to another.
Suggest why most of these blood transfusions failed.

....

.... **[2]**

(c) In 1865 Dr Lister introduced a method of improving the chances of successful surgery.
He used a carbolic acid spray as an antiseptic.

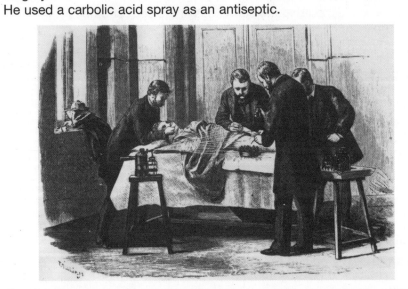

(i) Explain how this new technique increased the chances of successful surgery.

....

....

.... **[2]**

(ii) Suggest why the death rate from surgery was still high compared to modern surgery.

....

.... **[2]**

(d) In 1901, Dr Landsteiner realised the importance of different blood groups. The table contains information about the four different blood groups.

Blood group	Type of antigen on red blood cells	Type of antibody in plasma
A	A	anti B
B	B	anti B
AB	A and B	None
O	none	anti A and anti B

Yousef needs a blood transfusion.
His blood is blood group O.

(i) Which blood group can he safely receive?
Explain your choice.

Blood group ....

Explanation ....

....

.... **[3]**

(ii) Yousef recovers from his illness.
He wants to donate some of his blood.
Which blood group(s) could safely receive his blood?

.... **[1]**

(iii) What will happen to the blood if a patient received the wrong blood type?

.... **[1]**

(iv) Doctors are now considering using artificial blood for blood transfusions.
Suggest why using natural blood may now be dangerous.

....

.... **[1]**

(Total 14 marks)

[turn over

3 Yoghurt is made in a number of stages.

Milk

↓

Heated to 90°C for 15 minutes

↓

Cooled to 37°C

↓

Yoghurt bacteria added
(*Lactobacillus* and *Streptococcus*)

↓

Bacteria respire anaerobically producing lactic acid

↓

Lactic acid causes some flavour, other flavourings added

↓

Yoghurt produced

↓

Yoghurt kept at 4°C

↓

Yoghurt eaten!

(a) Suggest **one** advantage of using milk to make yoghurt.

....

.... **[1]**

(b) Explain why the following stages are used.

  (i) Keeping milk at 90°C for 15 minutes.

  ....

  ....

  ....

  .... **[2]**

  (ii) Cooling the milk to 37°C.

  ....

  .... **[1]**

(c) The bacteria use anaerobic respiration.

  (i) Describe the process of anaerobic respiration.

  ....

  ....

  .... **[2]**

(ii) Explain what would happen if the bacteria used aerobic respiration.

....

....

.... **[2]**

(d) *Lactobacillus* and *Streptococcus* require slightly different pH levels.
Suggest why this is useful in yoghurt production.

....

....

....

.... **[2]**

(Total 10 marks)

4 In 1798 Dr Jenner found a way of protecting people from smallpox.

(a) This method is now called vaccination.
Which type of organism causes smallpox?

Choose from:

bacteria   fungi   protozoan   virus

.... **[1]**

[turn over

(b) What is meant by the word 'pathogen'?

.... **[1]**

(c) Dr Jenner's method gave immunity to the patient.
Describe how immunity works.
One mark is awarded for correct spelling, punctuation and grammar.

....

....

....

....

.... **[4]**

(d) Many countries did not introduce compulsory vaccination for various diseases
until about 1900.
Suggest why.

....

....

.... **[2]**

(e) Antibiotics are often prescribed by a doctor to treat bacterial infections.
Explain why it is important to fully complete the course of antibiotics.

....

.... **[2]**

(Total 10 marks)

5 Scientists investigated the effects of fertiliser on the yield of different types of wheat.

Amount of fertiliser kg per hectare	Wheat produced tonnes per hectare			
	Type A	Type B	Type C	Type D
0	6.8	6.1	6.6	7.2
10	6.0	6.6	6.0	6.3
20	5.8	6.8	6.2	6.8
30	5.6	7.5	6.4	6.7
40	5.9	8.3	6.1	6.6

(a) Why was one investigation carried out without any fertiliser?

.... **[1]**

(b) Which type of wheat should be grown if the farmer cannot afford any fertiliser?

.... **[1]**

(c) What is the pattern of results for wheat type B?

.... **[1]**

(d) Name **two** elements found in large amounts in fertilisers.

1 ....   2.. **[2]**

(e) A farmer wants to use large amounts of fertiliser on his fields but local people object.
Discuss this conflict of different interests.
One mark is awarded for a clear ordered answer.

....

....

....

....

....

....

.... **[5]**

(Total 10 marks)

(Overall total = 60 marks)

Letts

Answers to mock examination Paper 1

1 (a) Flagellum

Myonemes **[2]**

(b) Chloroplasts (carbohydrate granules are acceptable) **[1]**

(c) Nucleus **[1]**

EXAMINER'S TIP

You must be very careful to ensure the feature you choose is completely acceptable. For example, the outer thick coat is not really a plant or animal feature.

(d) (i) 15

4

2   1 wrong answer = 1 mark **[2]**

EXAMINER'S TIP

It is easy to make a mistake in counting in what appears to be a simple question. Many candidates would avoid this mistake if they used a tally system, crossing out the organisms as they are counted.

(ii) Fewer Euglena in dish 2 because no carbon dioxide

The Euglena survive because they will produce carbon dioxide from respiration

Very low number of Euglena in dish 3 because of low (2°C) temperature/no light **[3]**

Answer written in complete sentences linking fact with explanation **[1]**

(iii) Equal numbers of Euglena to start with

Take more samples and take average

Repeat to check results

Leave for longer time

Investigate other variables such as pH, oxygen levels   any two **[2]**

(iv) No effect on animal numbers in dish two because lack of carbon dioxide unimportant

Slightly fewer animals in dish 3, lack of light unimportant, low temperature some effect **[2]**

2 (a) (i) vein **[1]**

(ii) Flaps in valves

Preventing back flow of blood so blood collects at valve **[2]**

(b) (i) Muscles in heart contracting **[1]**

(ii) Artery **[1]**

(iii) Since blood carries oxygen/dissolved food/waste carbon dioxide

It must be forced all around the body **[2]**

(c) (i) Blood in vena cava contains more carbon dioxide/nitrogenous waste than pulmonary vein

Blood in pulmonary vein contains more oxygen than in vena cava **[2]**

EXAMINER'S TIP

Did you remember to write a comparison? You must write down the situation in both blood vessels to explain a difference.

(ii) Pressure will be equalised in both ventricles

The left ventricle should have a higher pressure to force blood all around the body

Oxygenated and deoxygenated blood will mix

So blood leaving in the aorta will have less oxygen/lower pressure

So baby's cells/muscles are short of oxygen (Blue baby)   any three points **[3]**

EXAMINER'S TIP

If you had drawn arrows to show the direction of blood flow it would have helped you to work out some answers. Do not be afraid to mark or annotate diagrams if it will help you.

(iii) Advantages:

no decision made at difficult time for the family

potential donor may be incapable of deciding at that time

solve donor shortage

Disadvantages:

religious objections

personal freedom objections

unsuitable donors, e.g. Aids

would it be unenforceable?

what 'punishment' for not carrying cards?

2 marks for advantages
2 marks for disadvantages
1 mark for use of complete sentences
1 mark for logical argument **[6]**

3(a) A D E B

A before D = 1, D before E = 1, E before B = 1 [3]

(b) amylase to starch

lipase to fat

protease to protein   1 or 2 incorrect = 1 [2]

starch to glucose

protein to amino acids

fat to fatty acids and glycerol

1 or 2 incorrect = 1 [2]

(c) (i) emulsification (not digestion)   [1]

(ii) larger surface area for enzyme action in digestion   [1]

4 (a) See graph

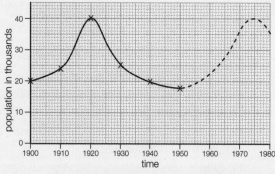

correct plotting = 2, one mistake = 1,
2 or more mistakes = 0
curved line through majority of points = 1 [3]

(b) (i) Mountain lions/wolves prey/eat Kaibab deer   [1]

(ii) Decreased

Overcrowding leading to lack of food/disease   [2]

(iii) Increase/population recovers

More food now available/more young survive   [2]

5 (a) 43

110   both answers required [1]

(b) Shout in pain involves the brain and thought
So it takes longer   [2]

(c) Reflex   [1]

(d) Receptor

Sensory

Electrical

Spinal cord   [4]

(e) (i) Chemical messengers/made in ductless glands   [1]

(ii) Hormones travel in blood

Not directly to where needed

Blood flows all around the body

any two points [2]

(iii) E.g. Insulin

Controls blood sugar   [2]
(Other answers possible)

6(a) 2   9

9   3   one mistake = 1 mark [2]

(b) Gases enter and leave

Named gas and reason, e.g. carbon dioxide

For photosynthesis/to make food

or

Oxygen

For respiration/to release energy

or

Waste product from photosynthesis
Control water loss
Preventing wilting/loss of turgor
Cooling effect
Of water evaporation   any four points [4]
linking of sentences [1]

(c) Leaf A (no mark for identifying leaf but answer must be correct)

Most stomata closed to prevent water loss/transpiration

Otherwise plant will wilt/carry out less photosynthesis   [2]

(d) (i) Less competition   **[1]**

For space

Water

Minerals   any two points **[2]**

(ii) Keeps atmosphere around plant

Moist /damp

So less transpiration/less water loss

Less water uptake since root hairs damaged
in transplanting   any three points **[3]**

7(a) (i) Normal/without Marfan's syndrome   **[1]**

   (ii) R   RR   Rr

      r   Rr   rr

           1 mark for correct gametes of parent

           1 mark for each correct line of offspring **[3]**

 (iii) rr/double recessive   **[1]**

 (iv) Mutations   **[1]**

(b) (i) Light in straight lines

Focussed by

Cornea/lens/fluid in eye   **[3]**

 (ii) Nerve impulses

In optic nerve/sensory neurones   **[2]**

 (iii) Lens cannot change shape

So focus cannot change/fixed focus   **[2]**

Now write in your scores for each question.

Question number	Possible score	Your score
1	14	
2	18	
3	9	
4	8	
5	13	
6	15	
7	13	
Totals	90	

Grading Zones		
Below 50 marks	Grade D	
50–60 marks	Grade C	Congratulations!
61–70 marks	Grade B	Well done.
71–80 marks	Grade A	Excellent result, you know and understand many aspects of Biology.
81 plus	Grade A*	Superb result, you have learnt a lot from Examiner's tips!

Answers to mock examination Paper 2

1(a) Oily feathers

Waterproof/insulate [2]

Thick layer of fat

Insulates/food store [2]

EXAMINER'S TIP

Only these two answers are acceptable since the other features do not affect the penguin's ability to survive cold conditions. Always read the question very carefully to avoid silly mistakes.

(b) Advantages – would find it difficult to keep more than one chick warm

difficult to find food for more than one chick

difficult to protect more than one chick

single egg can be bigger with more food

Disadvantages – if dies, parents must wait until next season

if a bad year population static

if a series of bad years whole population at risk

1 advantage and 1 disadvantage = 1 mark
2 advantages and 2 disadvantages = 2 marks
plus 1 mark for answer in complete sentences
and good English
plus 1 mark for correct linking of ideas and
comparisons [4]

EXAMINER'S TIP

To gain full credit, your answer must be well organised. Treat it like a question in an English examination by jotting down some facts, reorganising them into a logical sequence, then writing them out in complete sentences making sure your spellings are correct.

(c) Light

For flight [2]

(d) C

feathers/fewer digits/wing [2]

(e) (i) B A D [2]

B before A = 1, A before D = 1

(ii) B C A [2]

B before C = 1, C before A = 1

2 (a) Blood loss leads to tissue collapse

Less oxygen/food to cells

Loss of heat

Infection from wound from jaws of leech

any two [2]

(b) Different blood groups

Differences in pressure

Leakages

Transmission of disease

Air bubbles any two [2]

(c) (i) Killed bacteria

Which caused infection [2]

(ii) Ordinary clothes worn

No masks

No oxygen available

Spectators present

No modern drugs any two [2]

EXAMINER'S TIP

Do not be panicked by an unusual question! Remember it is still a Biology examination. So use your biological background to think of possible answers.

(d) (i) His own blood group O

Other groups have A or B antigen on red blood cells

Since group O has both anti A and anti B there is no serious reaction to either A or B antibody [3]

(ii) All groups [1]

(iii) Blood agglutinates/clots [1]

(iv) Possibility of carrying diseases such as AIDS/BSE/hepatitis [1]

3 (a) Uses excess milk/milk almost the 'perfect' food since it contains carbohydrates, fats and proteins [1]

(b) (i) Kill bacteria

So it does not go bad

(ii) Correct conditions for yoghurt bacteria [1]

(c) (i) Used in lack of oxygen

Produces lactic acid

Releases energy for growth/repair any two [2]

(ii) All sugars would be used up

No lactic acid for flavour

Less energy released for the bacteria

any two **[2]**

EXAMINER'S TIP

Many candidates show a lack of understanding when answering questions on aerobic and anaerobic respiration. The presence or lack of oxygen (not air) is the important factor. In anaerobic respiration the lack of oxygen results in only a partial breakdown of carbohydrates, so a different chemical (lactic acid or alcohol) is made which still contains some energy.

(d) Wider range of pH means reaction takes place more effectively/longer

Since the lactic acid produced causes low pH/ pH changes in yoghurt production **[2]**

4 (a) Virus **[1]**

(b) Disease causing

(c) Antigens

Stimulate white blood cells/killer T cells/ lymphocytes

To produce antibodies

Which react/destroy antigens

Difference between active and passive immunity

any 3 points plus 1 mark for well written answer **[4]**

EXAMINER'S TIP

Did you remember to plan your answer? Details of Dr Jenner's method are not required since you were asked to simply describe how immunity works.

(d) Technique not published

Lack of communications/poor spread of information

Resistance to new techniques

Cause of many diseases were not understood

any two points **[2]**

EXAMINER'S TIP

This question tests a new area at GCSE called Ideas and Evidence.

In your 'real' examination you may not be told which questions are based on this area since it will not affect your answer.

(e) Some resistant bacteria survive

These breed resulting in widespread immunity

So antibiotics have no effect in future

any two points **[2]**

5 (a) Control/for comparison **[1]**

('to find out the effect of no fertiliser' does not gain a mark)

(b) D **[1]**

(c) Yield increases as amount of fertiliser increases **[1]**

('yield increases' does not gain a mark)

(d) Nitrogen

Phosphorus

Potassium any two **[2]**

EXAMINER'S TIPS

Did you make the mistake of naming compounds such as nitrates and phosphates?
Always read the question carefully!
Other elements such as iron, magnesium are present in fertilisers but not in large amounts.

(e) Farmer – bigger yield

more profit **[2]**

Residents – fertilisers leached into rivers/lakes

rapid plant growth

death and decay of plants

lack of oxygen so fish and other animals die

called eutrophication any two points **[2]**

good linking of ideas **[1]**

Now write in your scores for each question.

Question number	Possible score	Your score
1	16	
2	14	
3	10	
4	10	
5	10	
Totals	60	

Grading Zones		
Below 35 marks	Grade D	
35–40 marks	Grade C	Congratulations!
41–45 marks	Grade B	Well done.
46–50 marks	Grade A	Excellent result.
51 plus	Grade A*	Superb result.

Index

Index